THE NEANDERTHAL'S NECKLACE

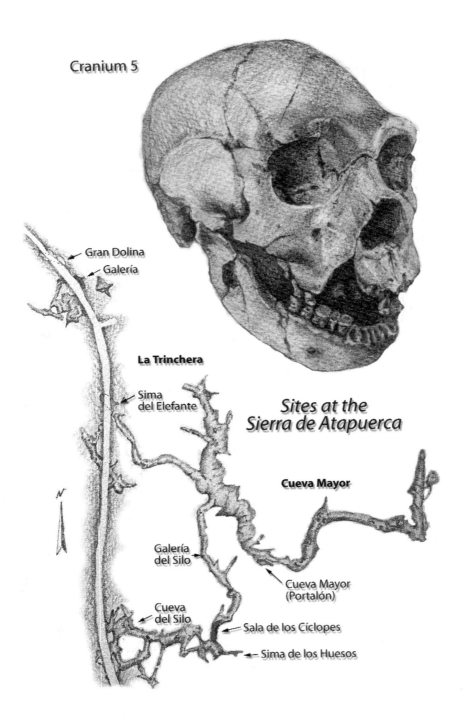

Cranium 5

Gran Dolina
Galería

La Trinchera

Sima
del Elefante

Sites at the
Sierra de Atapuerca

Cueva Mayor

Galería
del Silo

Cueva Mayor
(Portalón)

Cueva
del Silo

Sala de los Cíclopes

Sima de los Huesos

THE NEANDERTHAL'S NECKLACE

In Search of the First Thinkers

Juan Luis Arsuaga

Translated by Andy Klatt
Illustrations by Juan Carlos Sastre

WILEY

This edition published in 2003 by John Wiley & Sons Ltd,
The Atrium, Southern Gate, Chichester,
West Sussex PO19 8SQ, England
Telephone (+44) 1243 779777

Email (for orders and customer service enquiries): cs-books@wiley.co.uk
Visit our Home Page on www.wileyeurope.com or www.wiley.com

First published in the United States by Four Walls Eight Windows, New York in 2002

UNIVERSITY COLLEGE
WINCHESTER

0288985⁴ 599 938 ARS

Other Wiley Editorial Offices

John Wiley & Sons Inc., 111 River Street, Hoboken, NJ 07030, USA

Jossey-Bass, 989 Market Street, San Francisco, CA 94103-1741, USA

Wiley-VCH Verlag GmbH, Boschstr. 12, D-69469 Weinheim, Germany

John Wiley & Sons Australia Ltd, 33 Park Road, Milton, Queensland 4064, Australia

John Wiley & Sons (Asia) Pte Ltd, 2 Clementi Loop #02-01, Jin Xing Distripark, Singapore 129809

John Wiley & Sons Canada Ltd, 22 Worcester Road, Etobicoke, Ontario, Canada M9W 1L1

Wiley also publishes its books in a variety of electronic formats. Some content that appears in print may not be available in electronic books.

British Library Cataloguing in Publication Data

A catalogue record for this book is available from the British Library

ISBN 0-470-85157-0

Text design by Dr. Prepress, Indiana, USA.
Printed and bound in Great Britain by T.J. International, Padstow, Cornwall.
This book is printed on acid-free paper responsibly manufactured from sustainable forestry in which at least two trees are planted for each one used for paper production.

Table of Contents

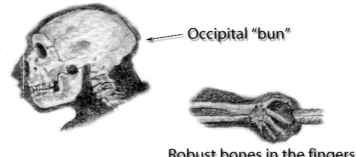

Occipital "bun"

Robust bones in the fingers

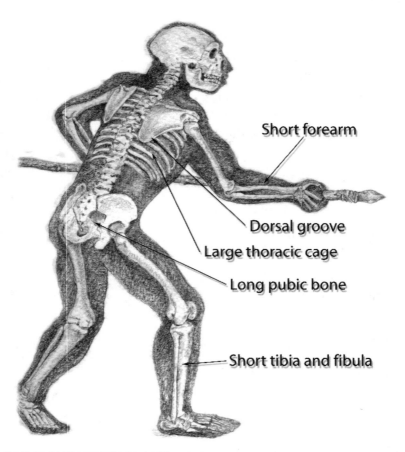

Short forearm

Dorsal groove

Large thoracic cage

Long pubic bone

Short tibia and fibula

Some of the features of the Neanderthal skeleton. (Based on Churchill, 1998)

Prologue

*To me those mountain mists are an indelible memory. I have for-
gotten other things. Feelings of affection and of animosity, acts of
kindness and expressions of disdain; these things are gone, leaving not a trace. But my spirit was transformed by those mists; they
reside within me now; never will they leave me.*

Pío Baroja, Fantasías Vascas

I look out the window at the rain. The drops of water sliding
down the pane seem incongruous, an intrusion of the natural
world on the artificial environment of concrete and asphalt that
is the city. Apart from us, there is scarcely a biological entity
within it. Though we are much more numerous now, we are the
same men and women who 25,000 years ago lived under the
open sky where large urban centers stand today. To be more precise, we are their descendants, the great grandchildren of those
hunters and gatherers of natural foods whom we imagine living
happily and in perfect harmony with the plants and animals of
the day. We think wistfully of the days when we lived as untrammeled a life as that of the American Indians on the movie screens
of our childhood. Free. Never having to go to the office.

People often ask me when I realized that paleoanthropology would be my vocation. Thinking back to find an answer, it comes to me that when I was a child, what I really wanted to be when I grew up was a hunter-gatherer. Maybe that is why I became a paleoanthropologist. All children are a little feral and untamed. We teach them things that are supposed to "civilize" them. We confine them within the four walls of a classroom. But somewhere within us all hides a prehistoric human who still responds to the call of the wild.

Of course we do not consider the appallingly high infant mortality rate of our prehistoric ancestors, almost half of whom died before the age of five. Nor do we think about the harsh winters with their endless snows, or famine in times of drought when the shadow of death spread mercilessly over small human communities. We imagine moments of sublime contentment, as when spring would arrive after a long winter and all that lived was reborn. We think about the fulfillment that we experience even during a brief foray into the natural world. By definition, nostalgia is a longing for *only the positive aspects* of the past. I recognize that there is much nostalgia in this book. But that is only the subjective aspect of the story.

By necessity this is also the story of formidable herbivores and powerful carnivores, of mountains and lakes, of the ice age and its glaciers, tundras, and taigas, and of Mediterranean forests. But it is also the story of smaller things, such as the falling of leaves in the autumn and the quiet footsteps of humans in the forest. The natural world is not a backdrop to the drama described herein. It is a protagonist. But fundamentally this is a narrative of our origins. It describes what we know about who we are and how we came to be so.

The book is divided into nine chapters and an epilogue. The first two chapters describe our place in relation to other living things—why we are so alone in the midst of so many species, why there is no other species on this planet that we can communicate with, who our nearest relatives were, and why they disappeared. The first few million years of human evolution are summarized, from the time in Africa before the emergence of a species that was able to populate first Asia and then Europe. If, as some others believe, our mental capacities were a recent evolutionary innovation, it would not be necessary to go so far back in time to trace their origins. However, if this thing called *mind* began its development long ago, when no human being had yet ventured out of Africa, then we have no option but to look to that continent not only for the sources of our physical structure, but also for the source of the human mind, our defining characteristic. In any event, the information provided to us by the first hominids sketches out a necessary background for our later consideration of another question: Apart from us, has there ever been another life form on earth that was conscious of its own existence and of its place in the world?

The third chapter considers the colonization of Europe and the glaciations that repeatedly covered much of the northern hemisphere over the last million years. The Neanderthals are discussed, as are their European ancestors, particularly those of the Sierra de Atapuerca, in what is today the Spanish province of Burgos. This concludes the first part of the book, fundamentally dedicated to the morphological modifications entailed in human evolution and the fossil record of the evolutionary process.

The fourth and fifth chapters focus on our ecosystems, or plant and animal communities, and how they have been transformed by one million years of European glaciations. This provides me the opportunity to discuss two of my great passions, mountains and forests. I am sure that many readers share my enthusiasm for what is left today of the natural world, but those who are not so interested in botany or in glaciers can skip these few pages without fear of losing the narrative thread of the book. I am confident that they will eventually return to this short digression when they want to know how the firs got to Cádiz or why the woods of Spain are so varied, even as seen from the windows of a passing car. The sixth chapter analyzes the role of humans in these ecosystems and the great wave of extinctions that took place when the ice melted and the present climatic epoch got under way. Wrapping up the most ecologically-oriented part of the book, we continue to the third and final section, in which we examine human mind and behavior.

The seventh chapter describes a remarkable paleontological site, the Sima de los Huesos, an ossuary in the Sierra de Atapuerca. Three hundred thousand years ago, more than thirty cadavers were deposited there by other humans in the earliest known funerary ritual. These people were conscious of the inevitability of death, a tragic discovery that changed life forever, depriving them and us of the blissful ignorance of the beasts of the field.

In the eighth chapter we explore the idea of consciousness and its inseparable companion, language. Are they detectable in the fossil record? When did symbols first appear?

With this information under our belts, we are prepared for the ninth and final chapter and the period of Neanderthal-Cro-Magnon coexistence, which ended with the disappearance of the Neanderthals. Fossil humans, climate, ecosystems, and the rugged Iberian terrain all have important roles to play in the final chapter. The narrative threads gather innumerable small events, occurring over a wide geographical area and an extended period of time. This story has long stimulated our imagination. It has given rise to a great deal of fiction, not all of it consistent with History with a capital H. I hope that it will be helpful to know which parts of those stories are realistic and which are implausible. This book provides some of the information that scientists are uncovering. With the benefit of that information, each person may envision events to his or her liking.

However, I would like to be up-front with the reader. Scientists now have a better idea when the Neanderthals disappeared, but it is not clear why or how they did so. Where science cannot provide information, speculation is the name of the game, because the circumstances of the events are subject to more than one interpretation. In this book I provide my version. The reader may arrive at different conclusions, since it is intuition and not reason that guides us in this mystery.

In any case, the Neanderthals are the primary protagonists of this book, not because they were our ancestors, but precisely because they were not. In the long evolutionary chain that links us to the first life forms billions of years ago, one link more or less would not have been very significant. But the Neanderthals were members of a parallel humankind that evolved in Europe independently from our evolutionary line over the course of hundreds of thousands of years. They provide a surprising mirror in which we can see people very much like ourselves but not quite the same. Thanks to the contrasts, we can understand ourselves better.

To make the reading of this book less of a chore, I have eliminated the use of acronyms to refer to fossils and the use of scientific names for present day plants and animals. The latter can be found readily in zoology and botany reference books. There is a summary bibliography of general works on paleoanthropology and prehistory at the end of the book, and a list of books and articles related to each chapter for those who wish to further explore specific parts of the story.

The goal of *The Neanderthal's Necklace* is to inform and provide the reader with some of the pleasure to be had in the daily struggle to answer the question that disquiets all of us at some point: What are we doing here? But I have a hidden agenda. Perhaps I should not confess it. I would like the reader to finish the book and come to the Sierra de Atapuerca, our sacred mountain, or climb the high lonely *páramos* of Ambrona, or examine the horses and bulls etched into the rocks next to the river and the ruins of the old water mill at Siega Verde. I would like the reader to visit any of the cave or rock shelter paintings on the Iberian Peninsula, or simply contemplate a mountain or a forest. In any of these places, I would like the reader to shiver with recognition, to feel exactly the same way I do when I see these things.

Two books on my desk have been an inspiration to me in the writing of this one. The first is *El Hombre Fósil*, written in 1916 by Hugo Obermaier (1877–1946), a great scholar of Spanish prehistory. This book was later expanded and translated into English as *Fossil Man in Spain*. I had the good fortune to purchase one of the rare surviving copies of a first edition run of two to three hundred from an antiquarian bookseller in the Netherlands. When I opened it, I found a handwritten letter from the author advising someone that a copy of the book would soon be sent. It was intended for a French-speaking colleague who was being addressed very respectfully, but whose name was not mentioned. The greeting of the letter is simply "*Cher Monsieur.*" From the marginal notations in the book I have deduced that it was none other than the famous paleoanthropologist Marcellin Boule, director of the Institute of Human Paleontology in Paris and Obermaier's boss, since the latter had been a researcher for that Institute since 1910. Obermaier was born in Germany but eventually took up Spanish citizenship. In the letter, he expressed his hope to meet Boule under better circumstances in the future. It was the time of the Great War, and Boule was in fact obliged to dismiss Obermaier from the Institute as an enemy national. This copy of Obermaier's monumental work, which contains his own handwriting on a letter and that of Boule in the margins, strongly evokes the history of human paleontology.

In *Fossil Man in Spain*, Obermaier successfully synthesized Spanish prehistory within the general framework of world prehistory. What is exemplary about the book is that it magnificently integrates archeological, geological, and paleontological knowledge. I could scarcely hope to achieve as much, and my book will be less formal, but I also intend to reflect the perspectives of the different disciplines involved in the excavation of a prehistoric site. What the field hath joined together, let no book put asunder.

The second book that I have on my desk, in the hope that its virtues will somehow rub off on me, is a work in two volumes, *Fisiografía del Solar Hispano* (*Hispania: The Physiography of Our Homeland*). It was written by Eduardo Hernández-Pacheco (1872–1965), the truly great geologist, naturalist, and prehistorian, and published in 1955. Hernández-Pacheco's books are remarkable not only for the depth of his scientific thinking, but also for the classic and elegant flavor of his writing, which seems to flow from the same Spanish earth whose properties he described. I consider *Don* Eduardo to be one of the twentieth century's great writers in the Castilian language, knowing as he did how to bring his beloved Spanish rocks to life for his readers. Interestingly, Herr or Don Hugo and Don Eduardo did not get along very well in life, but their books live in perfect harmony alongside each other on my desk.

In addition to these two classic works, I have another object on my desk. It is related to them in a sense, though it may not seem so at first glance. It is a copy of a sculpture, tiny really, of a woman's head with her hair gathered on top. The original was carved from ivory 25,000 years ago in Dolní Vestonice in

Moravia, a region of the Czech Republic. The head is very pretty, but to me it is more than a work of art. It is a manifestation of an exclusively human behavior, fruit of the capacity to communicate through symbols, to create language using images or sounds, to invent worlds and universes that may be fictional and even fantastic, but are just as real as reality itself. The books, the sculpture, and the computer I am writing on all spring from the same source. Creativity of mind and symbolic behavior, in general, form another of the important themes of the book you have in your hands, and one of the keys to understanding the demise of the Neanderthals and the reason for our absolute solitude today.

None of this is easy to express. I find it extraordinarily difficult, almost impossible, to translate into everyday language the lucubrations of researchers into the modern and prehistoric human mind. There are many books on the topic, but few are an easy read. I have to admit that at times I too find the psychological jargon to be excessive and artificial. Isn't there a simpler, more natural way to express these ideas? I think I have found the answer outside science, in the realm of metaphor. I was intrigued by a few lines by the great historian of religion, Mircea Eliade, that I found quoted in an article by Eduardo Martínez de Pisón. The excerpt opens the last chapter of this book. Mircea Eliade explains in those lines that the world "spoke" to "archaic man" in the days of societies with mythological underpinnings. The same metaphor flies like an arrow directly to the heart from the pen of Wenceslao Fernández Flórez in his book *El Bosque Animado* (*The Animated Forest*). I have taken the liberty of reproducing two excerpts from this moving book. I have also used the writings of other authors, starting with Shakespeare and Pío Baroja, to accompany my words. These quotes are not intended as decorations, but to serve as ambassadors for ideas. If those ideas are faulty, the ambassadors are not to blame. When all is said and done, poets and paleoanthropologists study the same phenomenon: human nature in its deepest and most mysterious sense.

At the end of the prologue in the Spanish edition of Obermaier's *Fossil Man in Spain*, he writes, "It is a fact that Spain possesses immense treasures related to fossil humans, and that the day will come when her studies of the Quaternary period will achieve a grandeur surpassing, perhaps, that of any other European country. This gives me great satisfaction. I am very enthusiastic about the future research of my friends and colleagues, and do not doubt that Chapter Six of this book ['The Iberian Peninsula During the Quaternary Period'] will itself be expanded into a large and splendid volume that may be entitled *Quaternary Spain*." Obermaier did not err in his prediction. Today the Iberian Peninsula occupies a very special place in European prehistory, as I hope to demonstrate in the pages to follow.

PART ONE

Shadows of the Past

CHAPTER ONE

The Solitary Species

The human as reconstructed by science today is an animal like the others—whose anatomy is so little separable from the anthropoids that the most modern classifications of zoology, going back to the position of Linnaeus, include the human with them in the same superfamily of hominoids. Yet judging from the biological results of the fact of the human being's appearance, is not the human precisely something entirely different?
        Pierre Teilhard de Chardin, The Human Phenomenon

So Similar, Yet So Different

We are unique and alone now in the world. There is no other animal species that truly resembles our own. A physical and mental chasm separates us from all other living creatures. There is no other bipedal mammal. No other mammal controls and uses fire, writes books, travels in space, paints portraits, or prays. This is not a question of degree. It is all or nothing; there is no semi-bipedal animal, none that makes only small fires, writes only short sentences, builds only rudimentary spaceships, draws just a little bit, or prays just occasionally.

The extraordinary originality of our species is not common in the living world. Most species belong to groups of similar ones. Thus we can observe a kind of continuity in the nature of similar species, broken only between the large groups of organisms. There is no intermediate form between the birds and the reptiles today, or between the reptiles and the mammals. Amphibians are not half-fish and half-reptile. Each of these different forms of vertebrate is traditionally assigned to a category called a class, with the exception of the fishes, which actually fall into three different classes: osseous fish, or fish with bones, the most common; cartilaginous fish like sharks and rays; and the lampreys, dramatically reduced in number today, but which were the first fish to appear. The vertebrates make up the majority of species among the chordates, which is a higher level category called a *phylum*. According to the hierarchy used by biologists to classify animals since the times of Linnaeus, the chordates occupy the highest phylum.

The chordates in turn are radically different from the various kinds of invertebrates like sponges, corals, the equinoderms, which is the group of sea urchins and starfish; annelids like earthworms; arthropods like insects, crustaceans, and spiders; mollusks like bivalves, snails, and octopi; and many other kinds, or *phyla* (the plural of phylum) of invertebrates. Each of these large categories is isolated from the rest with regard to morphology, which is the observable structure of the organism.

The ancient religious doctrine of the divine origin of species did not provide a satisfactory answer for the coexistence in the biosphere of species types that form clusters, which belong in turn to broader groups that exhibit enormous design differences among themselves. Was God a creator of such limited imagination that He or She could invent only a limited number of broad models from which He or She was obliged to develop variants?

The theory of evolution provides a different and more convincing answer to this problem: Similar species descend from a common ancestor that lived relatively recently, so they are closely related. The broader categories of organisms on the other hand, the phyla, were individuated long ago and have only remote common ancestors. After such a long period of independent evolution, it's logical that they are so dissimilar.

The first vertebrate fossils are more than 450 million years old, the first amphibians are more than 350 million years old, the first reptiles are more than 300 million years old, the first mammals are more than 220 million years old, and the first bird fossils are more than 150 million years old. Since the appearance of birds, though, evolution has not produced any really spectacular innovations. Could it have exhausted its inventiveness? To be honest, there is no exact method to decide when a group of species should be called a phylum or when it should be called a class or put into some lesser category. It is understood that a phylum is a broad category that describes an original biological design that is very distinct from any other organism in the same kingdom. A new phylum could develop at any time in the history of evolution. There is no reason to believe that important developments were limited to the remote past. Mammals are described as a class and do not constitute their own zoological phylum simply because there are other existing organisms with skeletons, and we are grouped together with them as chordates. But that does not mean that mammals aren't a truly original biological phenomenon. To a certain extent, the same can be said of us humans. The development of our intelligence really puts us in a new biological category. The French paleontologist and philosopher Pierre Teilhard de Chardin believed that we really deserved to be categorized as a new phylum.

So, if we are so different from the other mammals, does that mean that we have evolved separately for a long time? Not at all.

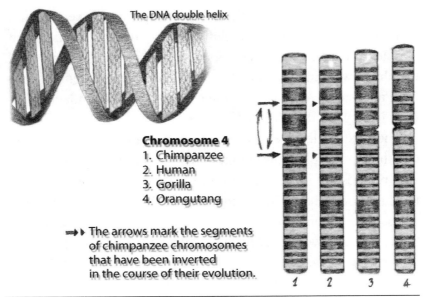

The DNA double helix

Chromosome 4
1. Chimpanzee
2. Human
3. Gorilla
4. Orangutang

➡▶ The arrows mark the segments
of chimpanzee chromosomes
that have been inverted
in the course of their evolution.

1   2   3   4

Figure 1: Chromosome 4. The genetic similarity between common chimpanzees, gorillas, orangutangs, and humans leaves no doubt as to the close relationship among all these species, although there has been an inversion in one part of the chimpanzee chromosome.

Our line is not one of the oldest by a long shot. It is no more than a measly five or six million years old. That was when our line and the line that would produce the chimpanzees diverged from each other. The gorillas' evolutionary line had diverged shortly before that. How then can we explain the dramatic difference between us and those other creatures? There are two parts to the answer. First, in some ways we have evolved very quickly. Second, all the intermediate forms between us and the chimpanzees have disappeared, along with their intermediate characteristics.

I began this chapter by noting some of the main differences between the human being and all other animals. Of those differences, only our upright posture is morphological. All the rest have to do with one particular organ of the body, the brain. Could it be that we are not very different from the chimpanzees after all? The fact is that we are distinguished from chimpanzees by only about 1% of our some thirty thousand genes. In fact, no more than fifty to one hundred genes are responsible for the cognitive differences between us and them. A very small but very significant genetic change has given us a unique intelligence, making us radically different from all other species and far from a mere variant of the chimpanzee. We are not just another species. Nevertheless, zoologists classify animal species according to their morphology and in recent years to their genes. So let's forget about our mental capacity for a moment and compare ourselves with other animal species from a morphological point of view. Let's go to the dissection theater and examine the bodies, *sans* brains, of various primate species.

Bodies Without Brains

The upper portion of Figure 2 shows who our closest relatives are. The closest is the chimpanzee, or rather the two existing species of chimpanzees. The gorilla is a little more distant, and then the orangutang. Within this group, the little gibbons are our most distant relatives, as anyone who has seen them in a zoo can easily understand. Chimpanzees, gorillas, and orangutangs resemble each other, and they are traditionally grouped together in the same family, the pongids. Gibbons are sometimes included in this pongid family, but some authors assign them their own family, the hilobatids. All the pongids, including the gibbons, are commonly called apes. Finally, the human species has the hominid family to itself. Together, the hominids and apes make up the hominoid superfamily.

This chart, which illustrates the evolutionary relationships among a set of primate species, is called a dendrogram, or tree diagram. Since they are all present-day species and no fossil species is included, this is not a species genealogy, or phylogram, and ancestor species are not named, although the ancestors common to two or more present-day species are represented by bifurcations, nodes A, B, C, D, and E. In the dendrogram, species converge in a specific sequence which represents the order in which successive divisions in evolutionary lines have occurred. The higher a node, the more recent the separation. In this case, the most recent division was the separation of the chimpanzees into two species (at node E) that have been separated by the Congo River for two and a half million years. The dendrogram does not communicate any other information, and it could be sketched out in many different ways without changing anything significant.

The dendrogram at the bottom of Figure 2 is actually the same as the one above, although the position of the human species has changed. Now it is not shown on one end of the hominoids, but *among* the others. This illustrates that the separate classification of the pongids and the hominids was artificial, since in reality the chimpanzees and the gorillas are more closely related to us than they are to the orangutangs and the gibbons. In other words, humans, chimpanzees, and gorillas have a common ancestor, a kind of "grandfather species" (C) that the orangutangs and gibbons did not descend from. The common ancestor of all the pongids, the hypothetical A and founder of the dynasty, was also our ancestor. If we were to be consistent, we would have to classify ourselves as pongids and as apes. The only logical alternative would be to classify all pongids as human. But would we then have to acknowledge their human rights?

All of this is interesting because it illustrates how evolutionary relationships and morphological similarity do not necessarily coincide. The chimpanzee is evolutionarily closer to the human but has a greater superficial resemblance to the gorilla

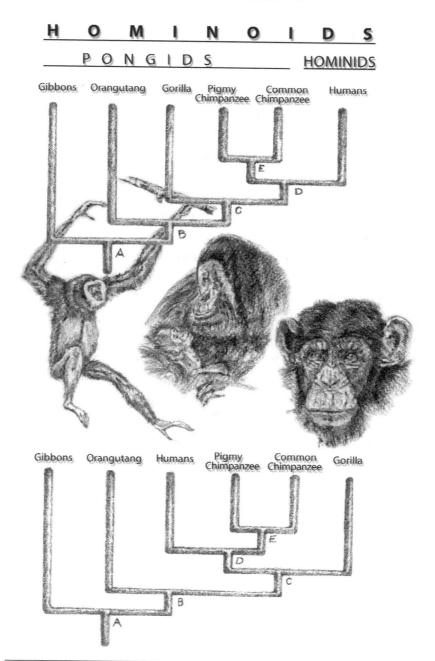

Figure 2: Two equivalent cladograms of the present-day hominoids.

and the orangutang. It was above all the German entomologist Willi Hennig who realized that we cannot rely exclusively on appearances to determine the evolutionary relationships between species. This discovery, seemingly so simple, was really brilliant, because it contradicted the apparent logic that similar-looking species should always be grouped together. Only great thinkers are able to see beyond the dominant paradigm to discover what the rest of us mortals cannot see, although it might be right before our eyes.

A similar case to that of the humans, although on a much larger scale, is that of the birds. This class includes many species whose closest relatives were the dinosaurs, specifically some small bipedal carnivores belonging to a group called the theropods. Other much larger and better known dinosaurs like Tyrannosaurus rex also belonged to this group. Actually, it would be more accurate to say that birds *are* dinosaurs, the only ones that survive today. There are big ones like the ostrich (and humans have encountered even larger ones), and small ones like the hummingbird. The birds were not the only feathered theropods, but only one part of a larger group. The velociraptors, made famous by the film *Jurassic Park*, probably had feathers, not scales as depicted in the movie. It's also possible that they were endothermic, or warm-blooded. Their feathers would have developed as an adaptation to help maintain a constant body temperature, since they have tremendous insulating qualities. The evidence presented to us by the birds is similar to that which would be available if all the mammals, save the bats, were to disappear. A hypothetical observer, long after the catastrophe, would be prone to imagine that all mammals had flown.

The birds were cut off from the rest of the vertebrates 65 million years ago, when a cataclysm wiped out the dinosaurs, or rather all the dinosaurs but the birds. Our own isolation is much more recent. "We" were still apes seven million years ago.

Actually "we" weren't anything, because at that time the two bifurcations that would produce the gorilla and human evolutionary lines had not yet taken place.

But if our study of bodies without brains illustrates our place among the primates, it also shows us how we differ from our nearest relatives, the chimpanzees. We are bipeds and they are quadrupeds, and our entire body and skeleton reflect this different form of locomotion.

Now let us consider fossil evidence to try to fill in the morphological gap that separates us from the chimpanzees.

Ape-men

The dendrogram in Figure 3 substitutes the scientific names, in Latin, for the everyday names of species: *Pan paniscus* and *Pan troglodytes* for the chimpanzees, *Gorilla gorilla* for the gorilla, and *Homo sapiens* for us.

Four new species now appear between the chimpanzee and the human: *Ardipithecus ramidus, Australopithecus anamensis, Australopithecus afarensis,* and *Australopithecus africanus.* None of them exist today, because they all disappeared more than two million years ago. All four were hominids, belonging to our evolutionary line, since all four developed after the bifurcation that separated our evolutionary destiny from that of the chimpanzees.

Notice that no fossil evidence for any other species of chimpanzee appears in the dendrogram. That's because we don't know of any. But there is no expectation that any chimpanzee's fossil will fill the gap that separates us from their live descendants. That's why the lack of that particular evidence is not important in this discussion. Nobody believes that there were chimpanzees in the past that were more bipedal or more intelligent than those living today. What is needed is evidence of what used to be called "the missing link," popularly known as an "ape-man."

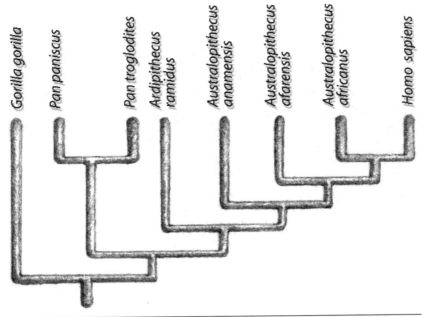

Figure 3: A cladogram that includes the Australopithecines.

Figure 3 is a kind of diagram that contains both fossil species and living ones, but where both appear at the same level, since it is neither a phylogram nor a species genealogy. No species appears as the ancestor of another. This dendrogram illustrates the different degrees of evolutionary relationships between and among species. Hominid fossil species have intentionally been placed between the chimpanzee and the human being. We've already seen that the location of a species to the left or the right on a dendrogram is irrelevant. What is important is how they are connected below. So the intermediate position of the hominid fossil species is purely arbitrary with regard to the phylogenic, or evolutionary relationships among the species. Nevertheless, the four hominid fossil species do represent the long sought-after "missing link," from a morphological point of view. Of course these links don't survive in any remote jungle or any other environment. They have been lost in time, where it is much more difficult to search for them.

To be sure, the only human characteristic that these fossil hominids, or most of them, had was an erect posture, a trait known as bipedalism. As far as the more outstanding of our characteristics, our brain, they were in the same category as today's chimpanzees. So they were hominids in the sense that they formed part of our zoological family, but they were not yet human.

'There are two species of alleged hominids prior to the *Ardipithecus ramidus* which, if they are truly hominids, will represent two new branches located to the right of the chimpanzees in our tree diagram.

The older of these two species is represented by a very complete cranium originating from Chad, in the centre of Africa, which today is a sandy desert but which in the fossil era, between six and seven million years ago, was a tropical jungle. The discovery of the species named *Sahelanthropus tchadensis* was made by the team led by Frenchman Michel Brunet in July 2001 and announced a year later. Not all authors agree that this is the cranium of a hominid, and it will not be easy to demonstrate whether it is or not. The reason for this is that because of its age, the fossil from Chad dates from a point very close to the time of separation of the evolutionary lines leading to the two types of chimpanzees, on the one hand, and to the human species, on the other; also the gorilla branch, which broke away earlier, cannot not be much older either. So, unless anything which remains of that geological age has a very clear human feature, such as bipedal posture for example, it will be difficult to prove that it is a hominid. Judging from the published photographs, I would bet that the Chad fossil was not a biped (although I admit that it is a risky bet). The cranium has a small face and canines, which seem to indicate that it is a female, but it nevertheless has a very robust supraorbital torus, which would indicate a male with a small face and canines. Just as we humans differ from our closest relatives, the great apes—where in both sexes the face is small and the canines very much reduced in size—in the Chad cranium we could have an ancestor with our facial morphology. On the other hand, such a developed supraorbital torus does not seem typical of a hominid. In any case, the controversy surrounding the significance of this cranium has only just begun, and we have yet to see what new discoveries will be made in that region of Africa and in others at sites more than six million years old.

A species originating from the Tugen Hills, in Kenya, christened *Orrorin tugenensis* by Martin Pickford, Brigitte Senut and their colleagues, is six million years old. In my opinion, the mandible and the teeth found to date show no clear characteristics of a hominid. There is also part of a very robust humerus which indicates that the primate in question moved around a lot through the trees. But there are also remains of two femurs which give us plenty to think about. Although incomplete, they are very similar to those of the australopithecus, hominids of a later era which were, of course, bipeds. Here, too, we are awaiting more remains which will confirm or refute the bipedal posture of *Orrorin tugenensis*'.

Ardipithecus ramidus lived in what is now Ethiopia almost six million years ago (5.8 million years) until a little less than four and a half million years ago (4.4 million, to be more exact). Tim White's paleoanthropology team has discovered numerous fossils of this species in recent years, most of which are still being studied. The results of the research on the body skeleton, or the postcranial skeleton, are still not available, so any hypotheses about their form of locomotion are purely speculative. However, a report has been published of a foot phalanx (5.2 million years old), with a morphology 'that is consistent with and early form of terrestrial bipedality' but it has not yet been confirmed that this was the only type of ground movement. This was undoubtedly a very primitive hominid that lived in the rainforest like today's gorillas and chimpanzees. Their dentition tells us that they had the same diet as the chimpanzees: fruit, tender stems, fresh leaves, and shoots. From this we deduce that they spent a lot of their time up in the trees, eating and sleeping. We still do not know how they got from tree to tree. However, the *Ardipithecus ramidus* had one characteristic that ties them to humans rather than chimpanzees. Their canines had begun to grow smaller. This single trait places *Ardipithecus ramidus* among the hominids.

The reader may find it somewhat confusing that I use dendrograms to illustrate human evolution instead of the more familiar family tree of species, or phylogram. I hasten therefore to advise you that you will find a phylogram in another chapter, my favorite in fact. But please don't skip ahead to see it. Actually, the dendrograms used here are of a special kind called cladograms. They follow a set of principles known as cladistics, laid out by Willi Hennig. The species are assigned to natural groups, or clades.

According to this school of thought, it cannot be known if one fossil species is the ancestor of another, be it fossil or living, because no one can travel to the past and thereby trace the course of evolution. All that can be scientifically established is the degree of kinship between species, and that is precisely what the cladogram reflects. Cladists consider hypothetical evolutionary trees to be purely speculative and lacking in scientific rigor. This is not to say that cladists deny evolution. On the contrary, they consider themselves to be the most faithful Darwinists. They simply will not go any further than to construct cladograms. Their cladograms are based exclusively on morphological information, so they are not affected whether a species exists or is a fossil, or if it comes from Africa or Australia.

I'm not a fanatical cladist, and I do believe that the age and geographical origin of fossils should be taken into account in our analysis of human evolution. When this additional information is added to that provided by a cladogram, we are able to construct what is called an "evolutionary scenario." This is an evolutionary narrative applied to a particular group that takes all available evidence into account, which is what I will try to do in this book. In our case, we have the advantage of an archeological record that is very helpful in the construction of a narrative. We also know some things about the hominids themselves and the climate and other elements of the ecosystem in which they have lived.

Although the accuracy of an evolutionary narrative as such cannot be proven, the elements upon which it is based can be confirmed or refuted by new evidence. If subsequent discoveries produce numerous and substantial changes in our understanding of the facts, someone may need to write a different book. We will have a few years to consider that possibility, maybe very few at the rate that paleoanthropologists are currently working. For now, we can safely say that *Ardipithecus ramidus* was a very ancient and primitive hominid of East Africa and that we are descended from that species or another very similar one that lived some 4.5 million years ago in Africa, probably in East Africa. We will soon know much more about the characteristics of the species; we just need a little patience.

Australopithecus anamensis is the next fossil hominid. We have a handful of their fossil remains that come from the Lake Turkana basin in Kenya, where they were discovered by Meave Leakey's research team near Kanapoi on the western shore and

Allia Bay in the east. All of the Kanapoi fossils except for one jawbone have recently been dated with great precision at between 4.17 million and 4.07 million years. *Australopithecus anamensis* had larger teeth with thicker enamel than *Ardipithecus ramidus*, which tells us that in addition to tender fruit, they also ate some other forms of vegetation that required more chewing, quickly wearing down the crowns of their teeth. These hard and abrasive plant products were probably grains and dry fruits. We think that they also ate the underground storage organs of plants, like bulbs, tubers, thick roots and rhizomes. The mineral particles that went into the mouth of *Australopithecus anamensis*, along with this food pulled from the ground, also helped grind down his teeth and contributed to their wear. All these vegetable foods are found in dryer forests than the rainforests inhabited by their presumed ancestors, the *Ardipithecus ramidus*. So it is thought that either *Australopithecus anamensis* had moved to another habitat, or more likely, their habitat itself had changed, becoming drier. A substantially complete tibia has also been found in Kanapoi, missing only its middle third, and its structure leads us to believe that these hominids had attained bipedalism. In the absence of another species from the same period, we can provisionally say that these were ancestors of ours. In any case, we descended from hominids similar to *Australopithecus anamensis*.

However, the *Ardipithecus ramidus* fossils are only two hundred thousand years older than the first fossils of this species, which provokes an interesting question. Would this have been enough time for such important anatomical and ecological changes to take place? Perhaps, for evolution does not occur at a constant rate. Sometimes it moves very quickly and then it seems to stand still. What is certain though, is that if we were to find 4.4 million-year-old *Australopithecus anamensis* fossils, *Ardipithecus ramidus* would no longer be considered our ancestor. It would be seen as a lateral branch of human evolution that did not lead anywhere.

Australopithecus afarensis, another hominid species, lived between 2.9 and four million years ago. Their remains have been found in Tanzania and especially in the Afar region of Ethiopia. Donald Johanson was largely responsible for these finds. We have a more extensive record of this primitive hominid species than of the previous ones, enough to tell us a lot about them. Dental evidence tells us that they too were completely vegetarian and lived in dry forests with clearings. They had an erect posture, but their arms were long in relation to their legs and they were still very good climbers.

Compared to our species, *Australopithecus afarensis* was small, barely larger than a chimpanzee. The males were probably about four foot five or just a little bit taller, and weighed about 100 pounds. The females were probably about three foot five, weighing sixty-five pounds or somewhat less. These estimates make the size difference between the sexes of *Australopithecus afarensis* greater than among us humans or among chimpanzees and closer to that of the gorillas. A male *Australopithecus afarensis* weighed about 1.5 times as much as a female. Among gorillas this ratio is about 1.6 times, among the common chimpanzees it is 1.3 times, and among humans it is 1.2 times.

The two most prized *Australopithecus afarensis* fossils are a very complete female skeleton nicknamed Lucy and an almost complete male cranium. The brain volume of this cranium has been estimated at just over 500 cc. Another less complete cranium seems sure to have had a capacity under 400 cc. This was apparently a hominid species with a brain barely larger than a chimpanzee's, whose brain size averages about 400 cc. Since their body weight was also similar to chimpanzees, we cannot attribute a much larger brain to the *Australopithecus afarensis* than to the latter even in relative terms, and they were probably not much more intelligent. The size of the human brain varies among our diverse populations and among individuals. Since it is a body organ, its size depends to a great degree on the size of the body. The average size of the human brain is usually said to be 1,350 cc, but our population is so large and so varied that this figure is more of a convention than anything else. In any case, it is interesting to note that the average brain volume of a human female is less than 1,300 cc while the average for males exceeds 1,400. As we will discuss in a moment, that does not mean that men are more intelligent than women. In addition, about ten percent of completely normal modern humans have a brain capacity of less than 1,100 or more than 1,600 cc.

As with the previous hominid species, we cannot be sure that *Australopithecus afarensis* was one of our direct ancestors. Some authors think so, but others do not, as we will see. This apparent confusion concerning human evolution may be unavoidable, and for a couple of reasons it is not as troubling as it may seem. In the first place, those seeking absolute truth or an immutable and unassailable dogma should look in a field other than science. In science one develops hypotheses, uncertain approximations of the truth that can always be wholly or partially modified by the weight of new evidence; but it is the best endeavor that the human spirit is capable of. Secondly, aside from the vanity of its discoverers, it does not matter so much whether or not *Australopithecus afarensis* is a member of our own evolutionary line. We can be reasonably certain that we have an ancestor essentially like *Australopithecus afarensis* who lived in Africa between three and four million years ago. That is what matters. Incidentally, prudence dictates that I say "in Africa" and not necessarily "in *East* Africa," as I would have said a few years ago, because Michel Brunet has encountered *Australopithecus* remains (those which in 1999 he called *Australopithecus bahrel-ghazali*) of the same age, around 3.5 million years, in Chad, Central Africa.

The next branch of our cladogram, moving towards the human being, is that of *Australopithecus africanus*. Their fossils have been dated at between three million and slightly less than 2.5 million years of age, and were discovered not in East Africa, but in South Africa, in three caves at Taung, Sterkfontein, and Makapansgat. They were physically similar to *Australopithecus afarensis*, and their brain does not seem to have been appreciably larger. The three best preserved crania, all from Member 4 (a specific level of excavation) at Sterkfontein, have the following capacities: 375 cc, 485 cc, and 515 cc.

The latter cranium seems to have housed a large brain, and some researchers have claimed that its cranial capacity must in fact have exceeded 600 cc. This specimen, like most fossils found in the South African caves, has been deformed by sedimentary pressure, so reconstruction is needed to correct its deformation. Glenn Conroy and other colleagues have used Computer Aided Tomography, a form of medical radiography, to examine the specimen. These CAT scans are being used more and more in the analysis of human fossils. The technique is to take a series of cross-sectional pictures very close to each other, as though cutting the fossil image into thin slices. A computer program then uses these two-dimensional images to reconstruct a three-dimensional view of the object. This three-dimensional view can then be manipulated on the computer screen, in this case to correct the deformation. Then it is possible to take measurements such as brain volume. This procedure has yielded a figure of between 500 cc and 530 cc, but some authors believe that this should best be considered a minimum.

In October 1998, the veteran paleoanthropologist Phillip Tobias, former director of excavations at Sterkfontein, and his lifelong collaborator Ron Clarke, the present director, announced the discovery of a very complete skeleton in Member 2, a deep excavation at the site. This skeleton could be as old or older than Lucy, who is 3.2 million years old, and it could even approach 3.5 million years. The circumstances surrounding the announcement of this discovery were bizarre, as things are from time to time in the world of paleoanthropology. In September 1994, Ron Clarke identified some elements of the left foot of this skeleton from among animal fossils that had been recovered at the site two years earlier. At that time the skeleton was dubbed Little Foot. Clarke and Tobias believed that it had very primitive characteristics, of the kind shared by chimpanzees and characteristic of an animal at least partially arboreal, although others disagreed. In May 1997, Ron Clarke found more remains of the same foot in the laboratory, as well as the lower ends of the left tibia and fibula, the lower end of the right tibia and a bone from the right foot, all from the same skeleton. Then, in June 1997, Clarke's on-site team was instructed to undertake a mission impossible. He told them to look on the walls of the big, deep, dark cave for a cut bone that would match the fragment of Little Foot's right tibia in the lab, something like looking for a needle in a haystack. And they found it in just two days! Little Foot was there, largely embedded in stone, as the walls of the Sterkfontein cave are a solid block formed of fossils in a hardened rock matrix. The cranium was complete. We will have to wait to know what hominid this was and to verify its great age and presumed arboreal characteristics. It will be an exciting wait. If this skeleton is of another species contemporary with *Australopithecus afarensis*, perhaps an old form of *Australopithecus africanus*, it will be a serious competitor for the position of ancestor of all subsequent hominids (that is, those less than three million years old), ourselves among them.

You Are What You Eat

Although we believe that the *Australopithecus africanus* lived in a similar environment to the *Australopithecus afarensis,* the former had larger teeth, indicating that their diet required more prolonged chewing. It was apparently based on plant material even tougher than that which their ancestors had eaten. But is there any way to find out exactly what the fossil hominids ate?

Any of the chemical elements can be present in various forms called isotopes. For example, our bones contain the isotope carbon twelve (C^{12}) as well as carbon isotope C^{13} in much smaller quantities. The difference between them is that C^{13}, or heavy carbon, has thirteen neutrons in its nucleus, and C^{12}, or light carbon, has twelve. Mat Sponheimer and Julia Lee-Thorp have done a wonderful study of the proportions between light and heavy carbon in the fossils found at Makapansgat, which are some three million years old and include representatives of the hominid *Australopithecus africanus.* In Africa, the trees and bushes retain less heavy carbon than do the grasses of open grazing lands, so the ungulates that consume grass accumulate proportionally more heavy carbon than those that eat the leaves of trees. These authors have analyzed the tooth enamel of *Australopithecus* and as expected, it contains less heavy carbon than that of the grazing animals like the reedbuck antelopes and the *Hipparion,* an extinct equid with three toes (in contrast to the single toe found on modern horses). However, the *Australopithecus* enamel contained more heavy carbon than that of forest dwellers like the spiral-horned kudu or the sitatunga antelope. Makapansgat

Australopithecines may have consumed the roots and seeds of high savannah grasses in addition to the fleshy fruits and tender leaves of trees, or they may have eaten insects that ate this grass or animals that grazed on it. They may have killed animal young or eaten carrion. The growth in molar size from *Australopithecus afarensis* to *Australopithecus africanus* makes me think more along the lines of grains, nuts, and underground storage organs than about animal products, though. To process animal products you don't need a larger chewing surface, but you do need instruments to cut meat and crush bones, which have never been found in association with these Australopithecines. In any case, the stable carbon isotopes seem to be telling us that the Makapansgat Australopithecines' habitat was not limited to dense forest. They also spent time in more open environments.

Almost Human

In the cladogram depicted in Figure 4, I've inserted two new branches between *Australopithecus* and us. Both represent members of our own genus. Like modern humans, they are *Homo*. The farthest from us and closest to the Australopithecines is a species called *Homo habilis*, the first of our genus. Evidence of their presence extends through Ethiopia (the Omo River Valley and the Hadar area), Kenya (Lake Turkana), and Tanzania (Olduvai Gorge). They lived from 2.3 million to 1.5 million years ago. It is interesting to note the geographical distribution of the last three hominid species that we have discussed, because if we want to say that *Homo habilis* descended from *Australopithecus africanus* and that *Australopithecus africanus* descended from *Australopithecus afarensis*, we would have to travel from East Africa (*Australopithecus afarensis*) to South Africa (*Australopithecus africanus*), and back again to East Africa

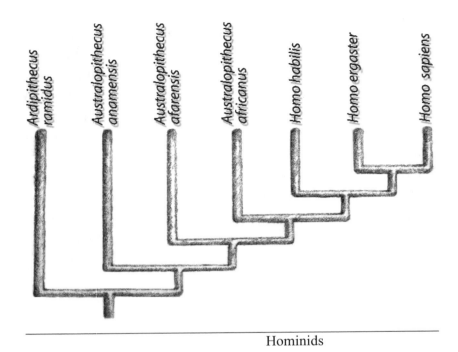

Hominids

Figure 4: Cladogram of the hominids. For the sake of simplicity, Paranthropus *is not included in the diagram.*

(*Homo habilis*). Biogeography significantly complicates our evolutionary scenario.

Some authors recognize fossils in existing South African collections as belonging to the species *Homo habilis*, saying that they could have evolved there from *Australopithecus africanus* and later extended their habitat to East Africa. I don't see *Homo habilis* in that evidence. Besides, the first *Homo habilis* fossils came from Ethiopia. My preferred hypothesis is that another hominid species descended from *Australopithecus africanus* and lived between three and 2.5 million years ago, and that *Homo habilis* in turn descended from that species. After putting the final touches on this book, I've had to return to this page because

Tim White's team has named a new species of hominid, *Australopithecus garhi,* on the basis of cranial and dental remains found on the Middle Awash River in the Afar region of Ethiopia, not far from the source of *Australopithecus ramidus.* These remains have been dated at about 2.5 million years and seem to fit into the evolutionary scenario to which I subscribe. It seems reasonable to think that the first *Homo habilis* developed somewhere in Africa from a hominid species like *Australopithecus africanus.* Other than Chad, East Africa, or South Africa, another plausible scenario would locate the process in Malawi, between those two regions. Tim Bromage and Friedemann Schrenk have found a jawbone about 2.5 million years old on the shore of Lake Malawi that they assigned to the genus *Homo,* although I'm not positive about the accuracy of that attribution.

Homo habilis was physically similar to the Australopithecines: short of stature with long arms and short legs. At least that is what our most complete skeleton seems to tell us. It is from Olduvai, Tanzania, where it was found by Donald Johanson and Tim White. Another partial skeleton discovered by Richard Leakey's team at Lake Turkana is similar. From a morphological point of view, there isn't much reason to accept *Homo habilis* as a member of our genus. It would seem more transparent to call the species *Australopithecus habilis.* Then the reader would better understand what kind of hominid we're talking about. However, it is not so clear that their intelligence was apelike, similar to that of the Australopithecines.

In the first place, *Homo habilis* had developed a somewhat larger brain than the Australopithecines. A *Homo habilis* cranium found at Lake Turkana has the smallest capacity, only 510 cc. This cranium actually isn't much different from some pertaining to the *Australopithecus africanus* of Sterkfontein. In some small and subtle, but important ways, it resembles us a little bit more. Another four *Homo habilis* crania have somewhat larger capacities: 582 cc, 594 cc, 638 cc, and 674 cc. The first is from Lake Turkana and the other three are from Olduvai. But there is a lot of doubt about the methods used to determine the volume of these fossil crania. They were very incomplete and/or deformed, and required extensive reconfiguration. We will probably see corrected estimates in the next few years, and they will probably be lower.

Brains themselves are not recovered, but their shape and dimensions are revealed by the endocranium, the cavity inside the cranium, which in life contained the cerebrum, the smaller cerebellum, and the medulla oblongata. Paleontologists fill the endocranial cavity with plaster, silicon, or latex, and mold a copy of the brain of a hominid that lived millions of years ago. It is the only body organ that survives in the fossil record, albeit as a negative copy.

To complicate the taxonomy, i.e. the classification into species, of the first fossils of the genus *Homo*, Richard Leakey found a cranium that doesn't easily jibe with our scenario. Richard Leakey has discovered many important fossils in the Lake Turkana region. He is the husband of Meave Leakey, of whom we have already spoken. Richard's parents, Louis and Mary, were the pioneers of human fossil recovery in East Africa, and had enormous success in Tanzania's Olduvai Gorge. The cranium that is going to complicate our story is called KNM-ER 1470, and its capacity is no less than 752 cc. It may have belonged to a large *Homo habilis* male, since this species exhibited a large degree of dimorphism, or morphological difference between the sexes, equal to or even greater than that of the Australopithecines. But this is difficult to believe. KNM-ER 1470 is not only larger than the other *Homo habilis* fossils, it is also structurally different. *Homo habilis* were like us in that they had much smaller faces and teeth than *Australopithecus africanus*. But in addition to its large brain, KNM-ER 1470 has an immense face and a large chewing apparatus, which is a strange combination of features. For this reason, some authors consider KNM-ER 1470 and some other large-toothed fossils like the above mentioned jawbone from Malawi to be a separate species, *Homo rudolfensis*.

Recently Maeve Leakey and other colleagues found on the western shore of Lake Turkana (Kenya) a cranium with a very broad, flat face dating from 3.5 million years ago, which they reported in 2001. It is contemporaneous with the *Australopithecus afarensis* of Tanzania and Ethiopia but its discoverers believe it to be a different species. The usual procedure would have been to give it its own species name which, in view of the flat shape of its face, would be *Australopithecus platyops*. But Maeve Leakey and her colleagues consider it to be the founding species of a branch of human evolution to which, more than a million and a half years later, the KNM-ER 1470 cranium from the other shore of Lake Turkana (dated at 1.9 million years) would belong. As it is a separate evolutionary line, it would therefore deserve a different generic name, and thus the new cranium from 3.5 million years ago would be called *Kenyanthropus platyops* and the KNM-ER1470 cranium *Kenyanthropus rudolfensis*.

Homo habilis also differed ecologically from earlier hominids. It was the first species not completely tied to a forest environment, be it the rain forest like *Ardipithecus ramidus* or the drier and less dense forests inhabited by the Australopithecines. *Homo habilis* seems to have inhabited much more open territory, like savannahs with trees and low vegetation either well dispersed or clustered between large expanses of grassland. This ecological change was of critical importance, because it opened the door to even greater changes that came later, changes that allowed the descendants of *Homo habilis* to live in every conceivable region, climate, and ecosystem. Without exception. every other member of our primate group, i.e. the gibbons, orangutangs, gorillas, and chimpanzees; and all our ancestors before *Homo habilis*, are or have been forest dwellers.

Homo habilis' change of habitat coincided with a dramatic climate change, a change that may in fact have caused it. Our planet has been growing continually colder and drier for four million years. Within the overall context of this tendency, there is also climatic fluctuation: a thermal oscillation that alternately heats and cools the earth, at the same time drying and dampening it. The thermal oscillation responds to astronomical factors such as the orientation of the earth's axis and the orbit that the planet describes around the sun. These astronomical changes follow certain cycles that affect the quantity and quality of solar radiation that arrives on the earth's surface, which along with other factors determine climatic cycles.

Until 2.8 million years ago, climatic oscillations were occurring approximately every 23,000 years, and were of low amplitude, i.e. they didn't cause dramatic changes. About 2.8 million years ago, however, the oscillations began to come only every 41,000 years and their amplitude increased considerably. Large masses of ice began to accumulate around both poles during cold periods. These arctic and antarctic icecaps may have become permanent at that time, surviving even the warmer periods, even if in reduced size. The periodic chilling and drying of the earth seems to have had an enormous ecological impact all over the planet, including in the African regions where the hominids lived. The rain forest shrank as open ecosystems expanded and encroached upon it. The expansion of the savannah and the resulting changes in vegetation were accompanied by the evolution of various lines of mammals adapting themselves to the new environment. *Homo habilis* was among them.

CHAPTER TWO

The Human Paradox

I have used the terms 'awareness' and 'consciousness' more or less interchangeably, although I tend to use 'awareness' (as in 'visual awareness') for some particular aspect of consciousness. Some philosophers make a distinction between them but there is no general agreement as to how such a distinction should be made. I must confess that in conversation I find I say 'consciousness' when I want to startle people and 'awareness' when I am trying not to.

Francis Crick, The Astonishing Hypothesis:
The Scientific Search for the Soul

The Invention

The adaptation of *Homo habilis* to the savannah, an open, grassy ecosystem, entailed more than a change of habitat. It was a change of ecological niche, the role that the species played in the chain of life, and the way its members made a living. For the first time, meat and animal fat were an important part of the hominid diet. Surprisingly, this change in the ecological niche doesn't seem to have caused any dramatic morphological transformation in *Homo habilis*, who as we have seen, still very much resembled the Australopithecines. Still, there were some minor changes in *Homo habilis'* head, resulting in a slightly smaller face and a somewhat larger brain.

This increased brain size may have had to do with a new way of life, now based on more dispersed and less predictable resources than in the tropical forest. This was true for the search for plant life and even more so for the hunt for meat. Their enlarged brains provided *Homo habilis* with additional capacity to make mental maps of large expanses of territory, to interpret animal tracks and other natural signs, like the flight patterns of carrion-eating birds circling a potential food source. They may also have been able to understand some of the rhythms of life and of the Earth, such as the changing seasons, enabling them to anticipate predictable events and plan for them. If they had this capacity, then it was a very significant change, because chimpanzees don't seem to make any type of plans for the future. It is also very likely that their social groups expanded and became more closely integrated and cooperative, their larger brains allowing for social and behavioral patterns that distinguished them from all other animals.

The primatologist Robin Dunbar has studied the size of primate brains and their constituent parts to see what variables correspond to the large brains found in many primate species. After ruling out a number of possibilities, Dunbar was left with only two hypotheses. Brain size corresponds either to a species' ecological niche or to the size and complexity of its social group. His final results demonstrated that there is a close relationship between the complexity of a primate's social group and the size of its neocortex, but no such relationship between the size of the neocortex and any ecological variable. The neocortex is the largest part of the human brain but is not the largest part of any reptile or non-primate mammalian brain.

The growth of *Homo habilis'* neocortex should thus be seen as a social phenomenon. Other hominids, specifically various species of *Paranthropus* (more about them later), also adapted to open ecosystems at the same time as *Homo habilis*, but their brain size did not greatly increase. Given that a larger neocortex affects mental functions such as associative and analytical capacity, I am certain that it enabled the first *Homo habilis* to develop their rich social medium and to occupy a completely new ecological niche. Their unusual social complexity may well have been the key to their ecological success, and later to the success of our species as well.

Finally, a great invention was produced. The morphological innovations that we have considered up to this point were brought about by evolution, the interplay between genetic forces like mutation and recombination on the one hand, and natural selection, or ecological forces, if we may call them that, on the other. But now intelligence had produced an innovation, qualifying it as the first invention, the flaked stone tool. The first reliably dated lithic (stone) artifacts ever found are from Gona, in the Hadar area of Ethiopia's Afar region. They are about 2.5 million years old. Other lithic assemblages found at Lake Turkana, at the Omo River, and in Congo, Uganda, and Malawi appear to be almost as old. The first human fossil associated with these artifacts is a jaw, a palate, and some teeth about 2.33 million years old, found by Donald Johanson's team, also in the Hadar area.

Judging by the abundance of grazing antelope fossils found here, the place was very open, clearly less wooded than it had been when impalas and *Australopithecus afarensis* had earlier inhabited the same region. The primate fossil, A.L. 666-1, is without a doubt *Homo*, but the attribution to the species *Homo habilis* of such an incomplete fossil is no more than a conjecture that I will allow myself. In any case, these artifacts at Hadar and other later ones associated with *Homo habilis* are crudely chipped stone tools and the sharp flakes that flew off the rock when it was struck to produce them. (Since it is not easy to say which of these were in fact tools and which were unusable by-products of the manufacturing process, some authors prefer to use the broader term *artifacts* to describe both the flakes and the stone core from which they had been chipped.) Archaeologists call this Oldowan industry, or Mode I technology. It has been said that the instruments produced by *Homo habilis*, whichever part of the artifacts they were, were "biological instruments." That is to say that their use augmented the morphology of the individuals who employed them. Since *Ardipithecus ramidus'* canines had begun to grow smaller, hominids did not have a good natural tool with which to cut the skin or bones of dead animals, nor any method to break open bones to extract marrow. Thus these stone instruments truly gave them access to a new food source.

Some animals can use carefully selected natural instruments, and chimpanzees can even modify them slightly to adapt to a desired function. They have been observed, for example, cracking nuts by using one stone as a hammer and another as an anvil. But no chimpanzee has ever been seen intentionally modifying a stone. Despite considerable effort, we haven't gotten them to carefully strike one stone against another to produce a cutting edge, even under experimental conditions. We can teach them well enough to use the edge of a stone flake to cut with, but they do not have the dexterity to produce one. All evidence indicates that their hands and arms are not well enough coordinated for such activity. Of course this is only a relative shortcoming, since they are so much more adept with their feet than we are. When chimpanzees want to intimidate someone though, they throw sticks or other objects very clumsily, in a fashion that bears no

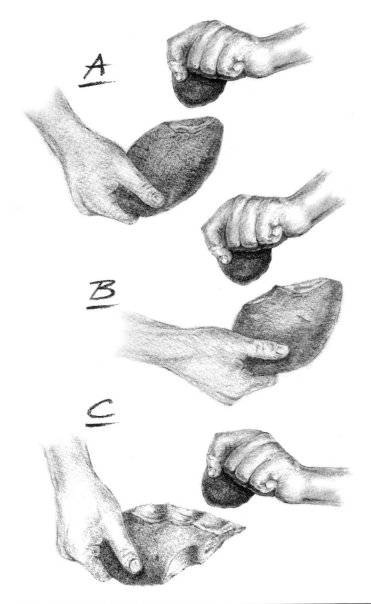

Figure 5: The chipping of stones. By striking one stone with another one or two times on the same surface (A and B), one can produce a cutting edge. If the chipping is continued and extended to two surfaces, a biface instrument is produced (C).

relationship to our concept of aim. Purposeful chipping at a stone is like sculpture in that it requires carefully chosen target points, very accurately aimed blows, a correctly calculated angle of impact, and well-regulated force. We should remember that under natural conditions chimpanzees are not presented with any situation that calls for the use of a cutting edge. Thus the process of natural selection has not favored the precise mental or anatomical characteristics required to produce one.

But the Australopithecines had arms and hands essentially like ours. They certainly had the mental and biomechanical capacity to produce tools, although none has been found in any of the sites where we found their fossils. Maybe they had no need for them. That's why I think that despite all the speculation around the appearance of the first tools, they in fact do not represent as momentous a mental leap as others have thought. In addition, this chipping of stone does not seem to reflect the existence of an ideal model tool, as though the toolmaker were working from a mental template of the desired instrument. He or she simply worked to produce a cutting edge of whatever kind emerged. Utility was the goal, not any specific configuration of the implement.

In any case, the cutting edges produced by *Homo habilis* ushered them into a new ecological niche, that of meat eaters. Perhaps we should classify *Homo habilis* only as "almost human," since their appearance represents more of a social and ecological step in human evolution than a cognitive one.

Be that as it may, the first toolmakers may have left evidence of conscious activities related to the use of their tools an incredible 2.5 million years ago. At some sites, the raw materials seem to have been transported from miles away for the purpose of manufacturing tools that would then be used in preparing animal meat. Of course no one has seen such activities in the remote past, but we have seen the evidence: abandoned stone tools with signs of their production by chipping and tool marks on the bones of herbivores, all in close proximity. One could speculate that the hominids would go look for stones when they saw a dead animal. But in regions where the necessary stones were rare, it is possible that the hominids brought some with them when they foraged, or that they even carried tools prepared in advance of their need. If they came upon some carrion, they may have needed to cut the meat and break open the bones immediately, for they would otherwise have run the risk of losing its meat and marrow to large predators or other carrion eaters. After all, competition would have been intense.

No tool-using animal, including the chimpanzee, shows such foresight. Instead, they look for a tool when they need one. Nor do they go far afield to find one. They just look for the materials they need within a radius of a few meters. Some day it may be demonstrated that the first hominids did not just produce their cutting edges in a mechanical fashion using whatever stones were available, or that they did not just crush bones with any large stone close at hand. If it can be demonstrated that they maintained the idea of acquiring appropriate stones over a period of time, then we will be able say that they exhibited conscious technological behavior more sophisticated than any that has ever been observed in a non-human species.

Animal behavior is intended to accomplish immediate and often visible goals. Chimpanzees sometimes hunt small animals, especially colobus monkeys, but they don't seem to have any strategy to lure or capture animals that are not in view and whose existence is therefore only hypothetical. They are opportunistic hunters, merely taking advantage of possibilities that present themselves. They do not form hunting parties. Instead, it seems that once a hunt has begun, other males simply join in.

Chimpanzees also make absorbent pads out of leaves to sop up liquids, and they peel twigs used to probe for termites and other insects; however, they always do so in order to use these tools immediately.

One may object that birds and beavers spend time carrying their nest-building and dike-building materials to their destinations. Ants too transport material to construct anthills, and we could list no end of animal building projects. Nevertheless, these are completely instinctive and preprogrammed behaviors. In no way do they indicate any perseverance of effort in order to achieve a consciously chosen goal. The perfectly hexagonal cells of a honeycomb do not in any way indicate the presence of architects among the bees. If it can be shown that the first hominids anticipated hypothetical carrion and therefore carried stone implements with them when they set out to forage, or even that they brought along stones to produce cutting edges if needed, I will be even more firmly convinced that they knew what they wanted to do and had a capacity for planning. How could such a thing be demonstrated? If we find a collection of herbivore bones either crushed or showing tool marks, but with no chipped stone at the site, it may be interpreted to mean that the hominids so prized their tools that they did not abandon them after use, perhaps due to a scarcity of raw materials. Tim White and his colleagues believe that they have encountered just such a circumstance at the same 2.5 million-year-old Ethiopian sites where the remains of *Australopithecus garhi* were recovered.

The First Human

The most well-known fossils of *Homo ergaster*, the next branch of the cladogram in Figure 4, are two crania and an almost complete skeleton that were found in the Lake Turkana basin of Kenya. The most complete cranium is 1.8 million years old and the other two fossils are 1.6 million years old. A team from the University of Florence has recently found a cranium in the Danakil Depression of Eritrea that preliminary studies assign to the same species. The age attributed to a new discovery is always tentative. Subsequent research is needed to firm it up. Nonetheless, these fossils seem to be only one million years old. If their species and age are confirmed, then the chronological range of *Homo ergaster* would be from something less than two million years to about one million years ago. Its geographical distribution would have been very wide, since the Swartkrans site in South Africa has yielded some fragmentary remains about 1.5 million years old that are believed to be of the same species. This would be the first time that we could identify one hominid species both in eastern and southern Africa.

Most *Homo habilis* fossils are 1.8 million to 1.9 million years old, although we may need to extend the chronological range of the species back 2.33 million years to the age of the Hadar fossil and the even less complete remains found on the Omo River. In principle, there is no chronological or biogeographical problem with saying that *Homo habilis* was an ancestor of *Homo ergaster*, whose fossil remains are 1.8 million years old or less. It is true that some specimens taken from Olduvai Bed II are commonly considered to be *Homo habilis* and are dated at 1.7 million or 1.8 million years old in one case and 1.5 million or 1.4 million years old in another. The rest of the *Homo habilis* fossils at Olduvai were taken from Bed I, the older level, and are about 1.8 million years old. There is also another fossil that I consider to be *Homo ergaster*. It is a coxa, or hipbone, that could be between 1.8 and two million years old. In any case, some authors believe that the most modern fossils found in Olduvai Bed II are too "evolved" to be *Homo habilis* and may be *Homo ergaster*. The stratigraphic origin and the age of the Lake Turkana coxa are not perfectly clear; they may be 1.8 million years old or even less.

To some, this possible (and only possible) chronological over-lap of the last *Homo habilis* fossils and the first *Homo ergaster* fossils makes it hard to accept the idea that the latter evolved from the former. In reality, a species' complete disappearance from the world does not necessarily have to coincide with the appearance of its descendant species in any given place. This would be a theoretical prerequisite only if one species evolved into another species throughout its entire geographical range, in a process that affected each and every one of its separate popu-lations. In most cases though, a descendant species evolves in a specific geographical location and from a specific population of its ancestral species. Thus the two may coexist over long periods of time within different geographical ranges. This was the case with *Homo ergaster* in the Lake Turkana region and *Homo habilis* in Olduvai. In fact, if a descendant species extends its range to other areas still inhabited by its ancestral species, the mother and daughter species could even coexist within one geo-graphical range. Eventually, if the two species occupy the same ecological niche, they compete with each other and the ancestral species could finally disappear. Broadly speaking, this is the mechanism believed to have produced the transition from *Homo habilis* to *Homo ergaster* and other similar sequences in the course of human evolution.

Homo ergaster was different in many important ways from all previous hominids. To begin with, their bodies had changed, reaching a size and a set of proportions similar to ours, very dif-ferent from the Australopithecines and the first *Homo* species. We know this from some isolated fossils such as the hip bone men-tioned above and a femur, both found on the eastern shore of Lake Turkana. But the development of our modern body size and shape in Africa by *Homo ergaster* was confirmed by the discovery by Richard Leakey's team of an extraordinarily well-preserved skeleton on the western shore of Lake Turkana, at Nariokotome. They found the skeleton of a boy who had died at the age of nine or ten and was as tall as, or taller than, a modern boy his age.

Homo ergaster had also developed significantly a greater cranial capacity, reaching 804 cc, 850 cc, and 900 cc in the best preserved crania. This increased brain size came about at the same time as an increase in body size and weight, so that in relative terms, it was not a great advance over *Homo habilis*. The cranial capacity of the latter may have been overestimated in the past.

Nonetheless, I find it hard to believe that the increased brain size of *Homo ergaster* relative to *Homo habilis* would not have meant a great leap in cognitive capacity as well. Whenever I have provided a figure for the brain size of a species up to this point, it has always been followed by information on its body size. That is because if a species increases its body size, one would expect that all its body organs would grow as well, and the brain is neither more nor less of an organ than the liver.

This scientific approach works well in general terms, when we compare diverse groups of mammals with very different body and brain sizes. At lower taxonomic levels it is less useful, though. To begin with, different people and different human populations have different brain weights relative to body weight, with no corresponding difference in intelligence. On average, and given equal body weight, a man's brain is 100 cc larger than a woman's. But this difference has nothing to do with cognitive, or "higher-level," mental functions. The evidence for this is that among macaque monkeys, who lack all such human intelligence, the sex-based disparity in brain size is in the very same proportion. It may be related to the capacity for visual and spatial information processing, which seems to be greater among males of our species and may be so among the macaques as well. In fact the clearest differences between men and women in psychological test results show up when they are asked to mentally rotate symbols, to remember the positions of things, to read maps, and to manage positional terminology. Dean Falk and other colleagues who have done this research on sexual dimorphism in brain size wonder whether natural selection has favored the capacity for spatial orientation in hominid males. Do men in general make better navigators? But enough speculation. Let's get back to our topic.

The brain of the common chimpanzee and of the gorilla are much closer in absolute size than are their body weights. The average chimpanzee brain weighs approximately 410 grams (about fourteen ounces), and that of the gorilla is approximately 500 grams (about 17.6 ounces). But the body weight of the chimpanzee averages about seventy-three pounds for females and ninety-five pounds for males, while average gorilla weights are 215 pounds for females and 353 pounds for males. So brain weight is proportionally much less in the gorilla, but as far as we know, they are no less intelligent.

In other words, two closely related species with similar brain size will usually have similar mental capacities (inherited from a common ancestor), regardless of disproportionate body sizes developed over time. The main reason for this is that the brain is an organ that consumes a lot of energy; it is expensive to maintain. Although our brains constitute only two percent of our body weight, they consume about twenty percent of the total energy available to our bodies, ten times more. But the chimpanzee's brain accounts for only nine percent of its total energy consumption. Given its high energy cost, we must conclude that if brain weight increases in the evolution of one species into another, there must be an important reason for it to do so. If there were no strong necessity for its growth, then natural selection would favor an unchanged brain size despite any increase in body weight.

Gorillas probably grew to their current size because they became folivorous, eating mostly leaves and stems, which are not very nutritious and can be processed only with the benefit of a large digestive apparatus. Conversely, the reduced body weight of a species will not make it more intelligent just because its brain size has remained constant.

In addition, the absolute size of the brain, not its relative size, is more meaningful with regard to some important aspects of physiology and behavior. I am referring specifically to the duration of individual development, which is longer for our species than for any other primate. This entails an extended period of nutritional dependency and other care before the individual reaches adulthood. The developmental period is also used to learn social interactions. These lessons are extremely valuable, since no member of a social species can survive alone, and ours is a highly social species.

Individual development among all the hominids before *Homo ergaster* lasted only as long as, or just a little bit longer than it does among chimpanzees today. In terms of skeletal ossification, our development is complete at about twenty years, although we usually stop growing in height a little earlier. Ossification is complete at about twelve or thirteen years among chimpanzees, gorillas, and orangutangs. Female chimpanzees have their first pregnancy at an average age of thirteen, sexual maturity generally coinciding with the end of bone growth. When the latter is complete, adult life and reproduction begin. When the female chimpanzee is growing, her weight increases day by day. Her diet provides her with more calories than she needs to survive, and she uses this surplus energy to build her own body. Once she attains reproductive maturity her weight no longer increases, but now she must utilize surplus calories to nourish a baby inside her body, expending energy to do so just as though she were supporting another bodily organ. Later, during the period of lactation, her dietary intake continues to nourish her child, now outside her body, until a new baby arrives. In a certain sense, a female is always "growing," first building her own body and then those of her offspring.

In the modern world, almost all human beings live in commodity societies where our plant and animal food is grown and raised rather than hunted and gathered. This way of life is in some sense artificial, especially in the wealthiest countries, where children enjoy an extraordinarily rich and varied diet. This may have affected some aspects of our developmental biology, for example the age of sexual maturity. In comparing humans with other species, therefore, it is best to use data from the few human populations that still eat like our ancestors, hunting wild animals and gathering wild plants. Among the Ache in Paraguay, for example, the first pregnancy occurs at about sixteen years and among the Dobe !Kung of Namibia at about eighteen, just about when the young women in each group stop growing taller. We will come back to these people and to pregnancy and children, but right now we should continue with our discussion of the brain.

A very close relationship has been demonstrated among primates as a whole between the size of the brain and the rhythm of the life cycle, i.e. the length of the different life stages. A long developmental stage provides an extended period for learning and preparation for adult life. That is why the length of childhood, adolescence, and life expectancy for a chimpanzee are double those for a macaque, whose brain is about one fourth the size of the chimpanzee's. Likewise, our life cycle is much longer than that of a chimpanzee. Our longevity is related to our large brain. Since the brain size of *Homo ergaster* was halfway between ours and the chimpanzee's, it makes sense to suppose that the duration of their childhood, adolescence, and life was also intermediate. At the age of three, a chimpanzee, a gorilla, and a human child all have a brain more than three quarters its adult size. This means that brain growth is mostly nourished by the mother, who provides the necessary energy, first during gestation and then during lactation. At the time of weaning, only a small part of brain growth is yet to come. So it makes sense to wonder about the purpose of such a long developmental stage. If I may employ a computer analogy, the answer is that this period is used in "programming" the brain, the installation of very complex "software" into our "hardware," already complete in its essential parts. For all the reasons that I have given, I consider the prolongation of this life stage to be very important, because I believe that it is a prerequisite for the development of a complex society and of increasingly elaborate technology.

Bifaces

A new type of stone instrument appeared in Africa 1.6 million years ago, and without a doubt, it was produced to be used as a tool. This was the biface, a large instrument chipped on two surfaces with obvious skill and concern for symmetry. There were various types, including hand axes, cleavers, and picks. The industrial stage to which they belong is called Acheulean, or Mode II technology, and reflects a significant technological advance over Mode I, the Oldowan, because there is a deliberate, that is to say conscious, intent to produce instruments of a predetermined design that existed only in the mind of its maker. Marcel Otte considers a biface, to all intents and purposes, a sculpture. Of course it was functional; it had utility. But it also reflected an aesthetic sense, a love of beauty. These primitive human beings were conscious of what they were doing, and they cared about the tools that they carried in their hands.

Based on their age, we believe that the bifaces were made by *Homo ergaster*. In fact, a *Homo ergaster* jawbone has been found together with Acheulean tools at a 1.4 million-year-old site in Konso, Ethiopia. In order to clarify the distinction between biological evolution and cultural change, it should be pointed out that the oldest *Homo ergaster* fossils are associated with Mode I technology. The invention and spread of Mode II technology did not reflect any biological change leading to increased intelligence. One and a half million years ago, some *Homo ergaster* populations were using a technology that was more elaborate than their ancestors had used, more elaborate than that used by other populations of the same species, and as we shall see, more elaborate than that used by some later humans. Although the appearance of a new technology does not necessarily suggest the development of a new species, we must also note that a very complex form of industry is not compatible with very limited

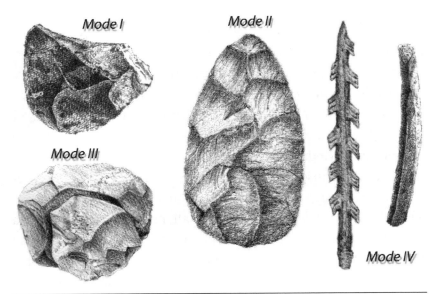

Figure 6: The four important technical modes.

intelligence. Some modern humans don't know anything about computers, but no monkey will ever learn to use one. It is unlikely, in other words, that *Homo habilis* had the intelligence to employ Mode II technology.

The famous French writer Jules Verne published a little-known novel in 1901 called *Village in the Treetops*. In it, he told the story of one French and one American adventurer who came upon some beings who appeared to be intermediate between chimpanzees and humans. These *Wagddis*, as they were called, lived in a village constructed atop the trees, in the canopy of an African rain forest. The novel perfectly reflects the spirit of those years of African exploration, a time when new lands, new animal species, and new peoples unknown to the West were being discovered with some regularity. These ape-men that Jules Verne invented represented the survival, in a remote and unexplored jungle in the Congo, of a missing link in human evolution. In one passage of the novel we read that "As to the microcephalic types, which it is claimed are intermediate between man and the ape, a species vainly predicted by the anthropologists and vainly sought for, that missing link which should attach the animal kingdom to that of man, could it be admitted that it was represented by these Waggdis?"

In *Homo ergaster* we find another "missing link between man and ape," to use the evolutionary jargon of the nineteenth century, although the "contact" between our species and *Homo ergaster* is taking place at a paleoanthropological dig rather than in a remote jungle. These human fossils are physically very similar to us from the neck down, but their brain size is not comparable to ours. Their intelligence must have been far short of ours, but far greater than that of the first known hominid, *Ardipithecus ramidus*, or that of the Australopithecines, and somewhat greater than that of *Homo habilis*. They were much more intelligent, to provide some context, than a modern chimpanzee. Their life cycle was longer at every stage, though still not as long as ours. They were capable of producing very elaborate tools that required a clear conception of the desired goal and of the many steps necessary in order to reach it. We are not talking about striking one stone against another two or three times in an arbitrary fashion.

The human being represents an extraordinary biological paradox that begins to manifest itself in all its enormity with *Homo ergaster*. Anatomically, we are but erect primates, an interesting evolutionary novelty, but in no way more extraordinary than the bat's capacity for flight or the cetacean's adaptation to the sea. At the same time, we humans are radically different from all other animals due to the astonishing phenomena of our intelligence, our capacity for reflection, and a broad self-consciousness of all aspects of our own behavior.

Homo ergaster was radically different from the first hominids in its ecological niche as well. They had definitively abandoned the heavily wooded ecosystems populated by *Ardipithecus ramidus* and the Australopithecines; instead, they lived off the resources of open environments. One of the world's most accomplished experts in prehistoric climates, Peter de Menocal, points to the African ecosystems from which *Homo ergaster* emerged as the result of a cooling and drying process precisely 1.7 million years ago that led to expanded savannahs and an even greater reduction of tropical forests. In addition to vegetation, *Homo ergaster* regularly consumed the meat of other animals obtained as carrion or by hunting. *Homo habilis* had already started down this road, but lacked the bulk needed for the hunting of large and medium-sized animals, therefore tending to eat more carrion than the hefty *Homo ergaster*. Finally, less than two million years ago, humans took a very important step. Their evolution, localized up until that time on one continent, became a worldwide phenomenon. The thinking bipeds spread beyond Africa.

An Extinct Branch?

But before leaving the African cradle of humankind I must touch briefly on the paranthropoids, another significant branch of the human evolutionary tree. These hominids were physically very similar to the Australopithecines, so much so that many authors include them in that category. They were also superficially similar to *Homo habilis*, who as far as we know did not represent any major morphological change either. The paranthropoids, though, boasted a notable specialization in their chewing apparatus, which was highly developed, enabling them to process large quantities of hard, fibrous, and very abrasive vegetal products.

Many researchers believe that the paranthropoids developed as a result of the same climate change of 2.8 million years ago that stimulated the origin of the genus *Homo*. This hypothesis is compatible with the paranthropoid chronology, because the first of their species, the *Paranthropus aethiopicus*, was in East Africa 2.6 million years ago. Later *Paranthropus* species were *Paranthropus boisei*, also in East Africa, and *Paranthropus robustus*, found in South African caves at Swartkrans, Kromdrai, and Drimolen.

One question that is not yet clear is where the *Paranthropus* species fit into the hominid evolutionary tree. The majority of authors place them at the level of *Australopithecus afarensis*, which would then be a common ancestor of *Paranthropus* and of the humans. This or something very similar could be correct. On the other hand, *Australopithecus afarensis* may not exactly be a common ancestor to *Paranthropus* and the humans, because to my colleague Ignacio Martínez and me, they appear in some respects to be "kind of paranthropoid" themselves. In that case, *Australopithecus anamensis*, or as I always say, "some very similar species" would be the common ancestor of the paranthropoid and human branches. To tell the truth, as far as we know, *Australopithecus afarensis* and *Australopithecus anamensis* are also two very similar species.

Populating Asia

Up to this point this has been an exclusively African story, like that of the gorillas and chimpanzees. Neither *Ardipithecus ramidus*, the Australopithecines, nor *Homo habilis* ever encountered any plant, animal, or natural landscape other than those they knew in tropical Africa, our first home. Every animal species is adapted to its ecological niche, the role it plays in the ecosystem to which it belongs. Evolution supplies each species with the tools it needs to occupy that niche, providing for its morphology (the structure of an organism, the part we can see), its physiology (the functioning of its body systems), and its behavior. Some species are environmentally intolerant. They can live only under very stable conditions, a kind of ecological bubble that constitutes their universe. It is very difficult for such species to escape their habitat, so their geographical distribution generally coincides with its boundaries. Gorillas and chimpanzees, for example, live only within the rain forest. As far as they are concerned, its boundaries are also the end of the world.

Other animal species are more ecologically flexible, more tolerant of environmental variation. Such variation may occur in the physical medium, such as the climate or the level of salinity, in the case of aquatic creatures; or it may refer to the biological medium, meaning the biological communities, or biocenoses, that they can successfully integrate into. Naturally, such flexible species have much wider areas of distribution. Sometimes evolution produces a species that occupies a new ecological niche, either within the same ecosystem or in another. Such variations are usually minor, distinguishing a species just a bit from its nearest relatives. Ecological change is very rarely profound enough to give rise to a new evolutionary line that completely changes the framework within which a species has developed up to that point. Of course this does sometimes happen. After all, some mammals have adapted to a life spent wholly in the sea, and others spend almost all of their time in the air.

The hominid species previous to *Homo ergaster* were without exception restricted as to habitat. Some of the first hominids lived in the African forest, and later species lived on the African savannah. But to reach East Asia, as we shall see, represents a great ecological change, and to colonize Europe, home to no other primate, an even greater change. Nevertheless, human adaptation to new and varied ecological situations outside Africa did not produce any spectacular morphological or physiological modifications. Only one organ made such ecological flexibility possible, the human brain.

The first East Asian hominid fossils were found in Java and China. Although Java is now an island, humans arrived there on foot. The explanation for this apparent paradox is that the islands of Java, Sumatra, and Borneo form an emergent island system situated on the Sunda Shelf, a large expanse of very shallow seabeds in the Indonesian Sea. When periods of widespread glaciation caused the world sea level to drop, the shelf itself emerged to connect the island system with the continent.

Human fossils have been found in several areas of Java. Unfortunately, we don't know the exact source of most of them, which makes it difficult to establish their geological age. In recent years, a team of geochronologists headed by Carl Swisher from the United States has been trying to establish a timeline for the fossil record of Javanese human evolution. Though we are hampered by the problem of indeterminate origins, it is believed that the oldest fossils are the following: a) The top of a child's cranium without a face, a *calvarium*. Found at Modjokerto, it is dated (or at least the sediment where we think it came from is dated) at 1.8 million years; b) Some incomplete and very deformed cranial remains found near Sangiran, dated at 1.6 million years. Again, this is actually the age of site that the fossils supposedly come from. The area of Sangiran, which is a large sedimentary depression, has in fact yielded a significant additional number of human crania.

These very old Javanese fossils are contemporary with the first *Homo ergaster* fossils in Africa and could be of the same species. The problem is that the child at Modjokerto died at three to five years of age, too young for us to be able to identify what kind of human he or she was, and the well-dated fossils from Sangiran tell us little. The truth is that we don't know much about the first humans in Java or when they arrived.

In 1891, Eugène Dubois found a skullcap, a tooth, and a femur at Trinil, another Javanese site. Based on these finds, Dubois proposed the species *Pithecanthropus erectus*. Today the species is called *Homo erectus*, but the original name reflected the then-current belief that *Pithecanthropus erectus* was a species of "ape-man," a creature that stood upright and resembled us in body but not in mind. In fact, Dubois was not so far off the mark.

The cranial capacities of all these Javanese fossils are not much greater than those of the *Homo ergaster* fossils. They vary from 813 cc to 1,059 cc. Recent studies by Carl Swisher have indicated that all the fossils from the depression at Sangiran are more than a million years old, even if their exact origins cannot be ascertained. Based on the animal fossils they are associated with, the human fossils at Trinil are probably contemporary with the most modern fossils from Sangiran.

Many authors believe that no distinction should be made between *Homo ergaster* and *Homo erectus* and that they should all be called *Homo erectus*. In such cases, the older name always prevails, and *Homo erectus* dates from the nineteenth century. I myself have studied the two species' cranial architecture and have arrived at the conclusion that there is no significant difference between them. Nevertheless, the Asian crania are more robust than the African ones, in some cases extraordinarily so. There are also other differences at the base of the cranium that could justify drawing a distinction between the species, so technical discussions among specialists continue. Yet these Asian and African fossils represent essentially the same type of hominid. For the purposes of this book, I could have used the term *Homo erectus* without reservation to describe the *Homo ergaster* fossils from Lake Turkana. In 2002, the international team of researchers working in the Middle Awash region, in Ethiopia reported a calvaria from the Bouri area dating from one million years ago. The authors attribute it to *Homo erectus*, but they are using this name in a wide sense which includes Asian and African fossils.'

Continental Asia was undoubtedly populated by humans before Java, because they would have had to walk across the former to reach the latter. There is a site in China, at the Cave of the Dragon in Longgupo, that may be the source of some stone tools and two human fossils almost two million years old, but this is not certain. The supposed tools are questionable, and one of the human fossils, a jaw fragment with two teeth, may in fact be from an orangutang. There is also some doubt whether the other fossil, while unquestionably a human incisor, is as old as the rest of the finds.

At Dmanisi (Georgia), sensational discoveries are being made of hominid remains from 1.8 million years ago. To date, four crania and three mandibles have been recovered. The team working with Georgian David Lordkipanidze has created a new species name for these remains, *Homo georgicus*. In China, a cranium has been found at Gongwaling and a jaw at Chenjiawo, each of which seems to be about one million years old. The cranium is very deformed and incomplete, but from what can be guessed it seems to be *Homo erectus*, the same species that populated Java at the time. But the best Asian collection of *Homo erectus* crania, or calvaria actually, comes from Zhoukoudian Cave, near Beijing. The age of these fossils seems to span an extended period between 300,000 and 600,000 years, although some authors think it is a little more and some say it is less. Franz Weidenreich, the scientist who studied them, was able to estimate the cranial capacities of five fossils with considerable exactitude. They ranged from 915 cc to 1,225 cc.

Beijing and Madrid are located at approximately the same latitude, close to 40° N, and the occupants of the cave lived in an environment pretty similar to their European contemporaries. We will discuss these ecosystems when the time comes, but let it be noted that conditions were quite different from those in their ancestral African environments or in the tropical forests of Java.

Practically all the human fossils from Zhoukoudian were lost in the Second World War, but new pieces of one of them, called Cranium 5, were found in subsequent excavations. This is an important fossil because it does not seem to have a larger capacity than the others, although it is the most modern of the group. A skullcap found at Hexian, with a cranial capacity not much greater than 1,000 cc, is another important *Homo erectus* fossil from China.

To date, only a few artifacts have been found in Java, and as is often the case with human fossils, their origin and age are problematic. The Ngebung site near Sangiran is a fortunate exception.

In recent years, a Franco-Indonesian team has found several stone tools associated with a human tooth in an unambiguous geological context. The tools are simple stone spheres and poly-hedrons, but we must take into account that the kinds of materi-al suitable for producing more refined implements are not found on Java. This is not the case in China. On the contrary, Zhoukoudian Cave has yielded thousands of artifacts. As in Java, they do not include bifaces, which has led scientists to believe that *Homo erectus* did not employ Acheulean technology. This may have been because they had no need for bifaces or because they left Africa before Mode II technology was invented there and never came into contact with the human populations that were employing it. A third hypothesis suggests that we need to reexamine our interpretation of the East Asian archeological record. The Chinese-American team of Huang Weiwen and Rick Potts have discovered 700,000 to 800,000-year-old tools in South China that they consider examples of Mode II technology, although they are not strictly Acheulean in design.

But while *Homo erectus* was getting established in East Asia, two other evolutionary lines were developing in Europe and Africa. At the end of this book, almost at the end of prehistory, these three far-flung actors meet again in a drama to be played out on a stage that includes most of the world outside of the Americas. The name of the European protagonist is Neanderthal man. In the next few chapters we try to familiarize ourselves with the Neanderthals' past, the days before the Great Contact with our ancestors from Africa. I have the good fortune to live on the Iberian Peninsula, where significant aspects of this history took place, and a fundamental part of the information now available to us comes from a dig in the Sierra de Atapuerca in Spain.

CHAPTER THREE

The Neanderthals

*And when the last remnants of this race are mouldering alone and
abandoned among the forests and the waters, a more fair-minded
generation will look upon the great plains and mountains of the
West, and will have to affirm, "Here lies the Indian race. It did not
achieve greatness because it was not permitted to."*

Karl May, Winnetou

Glaciation and Human Evolution in Europe

'All hominids dating from between 1.7 million and 5 million
years ago belong to a geological age known as the Pliocene, the
last age of the Tertiary period, within the Cenozoic age.
*Australopithecus anamensis, Australopithecus afarensis,
Australopithecus africanus,* and *Homo habilis* lived in the
Pliocene. But the oldest fossils of *Ardipithecus ramidus,* as well
as those of *Orrorin tugenesis* and *Sahelanthropus tchadensis*
belong to the Miocene, the age prior to the Pliocene.' The first
Homo ergaster probably lived before the end of the Pliocene, but
individuals less than 1.7 million years old lived in the next geological period, the Quaternary. Some fossils believed to be *Homo
habilis* are also Quaternarian. Early *Paranthropus* species lived

during the Pliocene, but *Paranthropus robustus* and *Paranthropus boisei* became extinct only during the Quaternary. All of these fossils are African, and we have already seen that it is difficult to identify the precise moment when humans populated Europe. It may have been before the end of the Pliocene or perhaps a little later. We can see that the transition from Tertiary to Quaternary coincides, by chance, with two important and possibly related moments in human evolution, the appearance of hominids that can justifiably be called human and the dispersion of these "bipedal apes" outside of Africa.

The Quaternary period is characterized by a general chilling of the planet, particularly in the last million years, during which there has been an intense cold period about every 100,000 years. These cold periods, called glaciations, have been particularly notable in the Northern Hemisphere. During glaciations, thick mantles of ice covered large parts of the boreal (northern) parts of Eurasia and North America. The greatest expansion of these glaciers took place about half a million years ago, when they covered all of Ireland, Scotland, and Wales, as well as all of Scandinavia. Their leading edges reached farther south in Europe than today's cities of Berlin, Warsaw, Moscow, or Kiev.

So much water froze during the glaciations that the sea level dropped to a point more than 325 feet lower than today. The glaciations alternated with warmer periods called interglacials. We are currently living in an interglacial, which provides the name for our epoch, the Holocene. The rest of the Quaternary is known as the Pleistocene, although for some authors it doesn't make much sense to consider the Holocene a separate epoch from the Pleistocene, since it covers only the last 10,000 years and is distinguished only by our own existence. But while human activity has drastically changed the biosphere and caused the extinction of many species, it has not yet produced any new species. So the terms Quaternary and Pleistocene are practically synonymous.

The existence of Quaternary glaciation was discovered through the study of the evidence left by glacial ice in its advances and retreats, including mountains as far south as the Sierra Nevada of Andalucía, and through the discovery of fossils of cold-weather species in sites where they could not survive today due to more temperate conditions. Later on, we will see that there were once reindeer on the Iberian Peninsula and that the mammoths found frozen in Siberia from time to time once ranged as far south as Granada. The waves of intense cold on the European continent are traditionally grouped into four periods established on the basis of the various levels of alluvial deposits, or terraces, left in the Alps during Alpine glaciations called the Günz, Mindel, Riss, and Würm, each named for a tributary of the Danube.

Within each of these glaciations we find very cold peaks called stadials separated by warmer phases called interstadials. The temperate periods between glaciations, like the one we have been enjoying for the last 10,000 years, are called interglacials. They are always longer and warmer than the interstadials. The glaciations have been studied in the Netherlands and Germany based on the observable geology of the North Sea basin. The four periods there are called the Menap, the Elster, the Saale, and the Weichsel, and the three interglacial periods are called the Cromer, the Holstein, and the Eem.

But the modern study of climatic change leads us directly to the ocean floor, where microorganisms called foraminifera are found among the other elements of marine plankton. Foraminifera are single-cell microcrustaceans with a calcareous shell, which is a shell made of calcium carbonate. When a foraminiferum dies, its shell sinks onto the seabed. Slowly but surely, the shells build up one on top of the other. But there are warm water foraminifera species and cold water foriminifera species. Thus we can probe the ocean floor and "read" the changes in water temperature over time in the succession of foraminifera species to be found there. Projects are under way to map a sequence of these microfossils over an extended period of time.

In addition, the proportion in seawater of two forms of oxygen (called "heavy" and "light" oxygen isotopes) varies in accordance with the general temperature of the planet. These chemical changes can also be tracked by analyzing the composition of foriminifera shells. By examining this temperature record on the ocean floor, we have been able to chart a very precise curve of paleotemperatures within which warm and cold phases can be delineated. These are the oxygen isotope stadials, otherwise known as OIS. They are numbered back in time from today's OIS 1. The odd numbered isotopes represent warm periods and the even numbered ones represent cold periods. In this book I will refer to modern isotope stadials and to classical glaciations, which are cycles that include various odd and even stadials.

One very direct way to discover the proportion between different oxygen isotopes in the past is to look directly at fossil ice. Researchers probing the Greenland icecap have reached 120,000-year-old ice, and the mantle of ice over Antarctica is now being explored. It is thicker than the ice in Greenland, and it is expected to reveal a very accurate record of the last half million years of climate change.

The Quaternary can be divided into four long stages in relation to human evolution in Europe. In the Lower Pleistocene, between 780,000 and 1.7 million years ago, the first humans arrived on the continent. The oldest fossils, about 800,000 years old, are from the Gran Dolina site in the Sierra de Atapuerca in Spain. For the time being we have only about eighty fossils, but excavation has been very limited to date. Since the fossils represent all parts of the skeleton and come from at least six individuals, it is likely that there are more human remains waiting down at Level 6 of Gran Dolina. We scientists who work in the Sierra de Atapuerca have an appointment to keep with them. But the excavations at higher levels must proceed at a cautious pace due to the meticulousness demanded by the scientific method, so I fear that it will still be a few years in coming.

We can conclude from the approximately eighty fossils we already have from Gran Dolina, that these people were neither *Homo ergaster* nor *Homo erectus*, but some other human species, more modern and evolutionarily closer to *Homo sapiens*. Their cranial capacity seems to have been greater than 1,000 cc. Since they are not of our species and they are not Neanderthals, we will need to find some specific name for them. One possibility is *Homo heidelbergensis*, but we think this name is more appropriate for the ancestors of the Neanderthals, so we have opted to call the human species that left their fossils at Gran Dolina *Homo antecessor*.

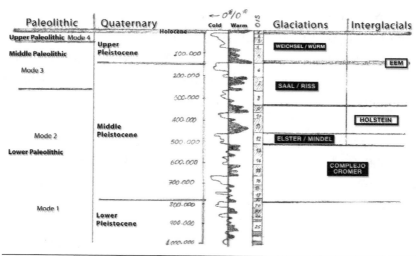

Figure 7: The Quaternary

From what we can tell at present, their characteristics situate them a little before the division of the evolutionary lines that would separately lead to the Neanderthals and to our species. I will leave it at that for now and for the sake of simplicity we can say that the Lower Pleistocene was "the period when people first populated Europe."

The people of Gran Dolina did not have biface tools within their repertoire. They produced only tools corresponding to the first of the great technical modes, Mode I. The vanguards of human migration, both in East Asia and in Western Europe, were still producing pre-Acheulean technology long after Mode II had been developed in Africa.

One important aspect of human behavior at Gran Dolina can be deduced from the fossils themselves. There is evidence among many of them that their flesh was cut and sliced from the bone by other humans without any deference to their common humanity. Jane Goodall has been able to establish that neighboring groups of chimpanzees in the Gombe Forest engage in violence so ferocious as to include the extermination of one or the other group, including occasional cases of infanticide and the consumption of the young victims. But the flesh of at least six people of all ages at Gran Dolina was put to very calculated, meticulous, and efficient use, as if they had been game animals captured for their meat, a chillingly inhuman act, or perhaps a chillingly *human* act, since we know of no similar behavior among other primates.

The ancestors of the Neanderthals lived and evolved from 127,000 to 780,000 years ago, in the Middle Pleistocene. The oldest fossil of the period is a jaw about half a million years old found at Mauer, near Heidelberg. The fossils from Sima de los Huesos, a cave site also in the Sierra de Atapuerca, stand out among the evidence of the Neanderthals' "grandparents," and they constitute the largest collection of human fossils not only in Europe, but in the whole world. The species name *Homo heidelbergensis* can be applied to all of them, in honor of the Mauer jaw. They practiced Acheulean industry, producing the bifaces characteristic of Mode II technology. Together, Mode I and Mode II form a cultural period classically known in prehistory as the Lower Paleolithic. The reader may find it somewhat odd that the Lower Paleolithic extends into both the Lower and Middle Pleistocene epochs, but the boundaries of cultural periods in prehistory do not coincide with the boundaries of the geological time scale. In fact, geological time is the same all around the globe, while archaeological periods are not simultaneous for all human species or populations.

The human type known as Neanderthal, known by the scientific name of *Homo neandertalsis,* took on a definitive identity in the Upper Pleistocene. Although they developed in Europe, the Neanderthals ventured out of this small continent to populate the Middle East and Southwest Asia. The entire epoch between 40,000 and 127,000 years ago is "the Neanderthal period." Neanderthal culture is characterized by Mousterian industry, corresponding to Mode III or Middle Paleolithic. If you are not sufficiently confused, let me add that the European Middle Paleolithic lasted from the next to the last glaciation, during the Middle Pleistocene, until about the middle of the most recent glaciation, during the Upper Pleistocene. I hope that the schema I am presenting helps the reader to orient himself or herself within this terminological jungle.

In any case, during the most recent glaciation, about 40,000 years ago or perhaps a bit more, some African immigrants appeared in the Iberian Peninsula and in Europe. They were our ancestors, the first European representatives of the species *Homo sapiens*, popularly known as Cro-Magnons. After a long period of coexistence, 10,000 years or more, the Neanderthals disappeared less than 30,000 years ago, just as the most severe phase of the last glaciation was beginning. Later on I will discuss this additional concurrence of significant events in climate change and human evolution. For now suffice it to say that we have been the only humans and the only hominids on the planet since that time.

We will call the twenty millennia from 10,000 to 30,000 years ago "the Cro-Magnon period." These modern humans arrived in Europe with their own Mode IV Upper Paleolithic technology. Some Neanderthals eventually used this technology as well. We do not know if they learned it by imitation or if they developed it independently and spontaneously. In time, Upper Paleolithic industry subdivided on our peninsula into a sequence of technologies, although they did not develop at exactly the same time in all regions. They were the Aurignacian, the Gravettian, the Solutrean, and the Magdalenian. In due time I will introduce another very important term, Chatelperronian.

The last 10,000 years constitute the Holocene, the interglacial in which we now live. Soon after the beginning of this latest epoch, cereal-grain agriculture took off in the Middle East, the earliest known sites being at Tell es Sultan in the Jordan Valley, also known as Jericho, and at Çatal Hüyük in Turkey. But it would take another two thousand years for agriculture and animal husbandry to arrive in Spain. This intermediate period of prehistory is called the Mesolithic.

The domestication of plants and animals resulted in what we call the Neolithic revolution, the cultural development that produced our fundamental economic conceptions and therefore our way of life. Since that event, humans have produced their food instead of harvesting it directly from the natural world through hunting, searching for carrion, or gathering naturally occurring plant foods. The new economy of food production entailed permanent human settlements, tied our species to the land, and led to tremendous population growth, so much population growth that we have utterly transformed the face of our planet. There are more than six billion of us now, still following the biblical mandate, "Be fruitful and multiply." The only food that we do not produce ourselves and that we still consume in great quantities is oceanic fish. But we are exhausting that resource as well.

We have not given any name to the interval of time between 30,000 and 40,000 years ago, the epoch after the Cro-Magnons had arrived in Europe and before the disappearance of the Neanderthals. Let's call it "the period of Neanderthal-Cro-Magnon coexistence." Although 10,000 years is very little time on the geologic scale, in human terms it is a long time, representing the entire duration of the Holocene.

The European Upper Pleistocene, then, is the period of both the Neanderthals and the modern humans who replaced them, although both human types have deeper roots in the European and African Middle Pleistocene, respectively. The Upper Pleistocene began 127,000 years ago with a climate as warm as that of today, although today's magnification of the greenhouse effect by gaseous industrial emissions has elevated temperatures beyond what would otherwise have been the case. The greenhouse effect in itself, apart from its intensification by human activities, is a natural phenomenon that is actually necessary for a large proportion of the life forms on the planet. The Upper Pleistocene began during a warm period that lasted some 10,000 years. The sea level rose as much as it has in our time. It was so warm that hippopotami returned to England, where they had, in fact, lived once before.

Then the weather changed. Colder times came, and with them the onset of another glaciation, the most recent. The earth's temperatures oscillated, as I will describe farther along, but reached

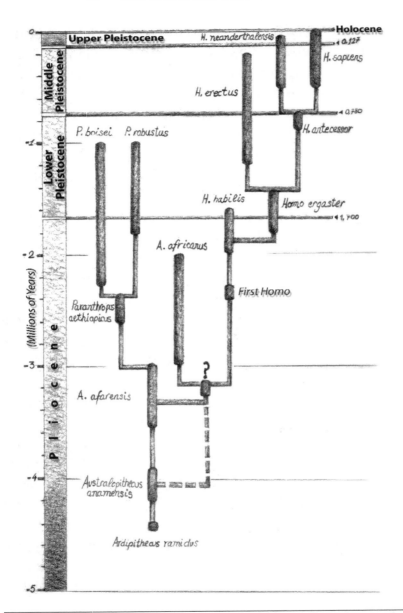

Figure 8: Human Phylogenesis. Australopithicus afarensis *may have been one of our ancestor species, or it may only have been an ancestor of the paranthropoids, in which case we should really call it* Paranthropus afarensis.

their lowest point between 17,000 and 21,000 years ago. By then there were no more Neanderthals. It was our ancestors who had to endure the long winters and deep snows.

However, the reward for that endurance was great. This was an ideal epoch for great herds of herbivores and their large hunters, both human and animal. After the cold years, the earth began to thaw, but shortly the cold returned for a period of less than 1,000 years. This short period is called the Younger Dryas. It marks the final stage, some 10,000 years ago, of the last glaciation. With the melting of huge volumes of snow and ice, large animals of prey moved away or disappeared, leaving some of their formidable predators to enter a period of decline.

— Last glaciation
··· Coastline
Elevated area: Maximum extent of ice.

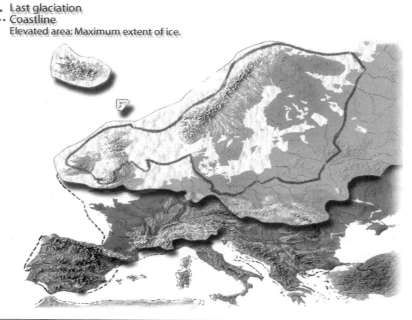

Figure 9: The mantle of glacial ice. In addition to the great masses of ice, mountain glaciers developed in the Alps and the Pyrenees (raised in the figure), and in many other European mountain chains. (Based on H. Kahlke, 1994)

The Calpeans

The Neanderthals were nearly called Calpeans, in honor of a Neanderthal-type cranium found at Forbes Quarry in Gibraltar in 1848. Calpe is the classical name for Gibraltar (There is another large rock and town called Calpe in Alicante, Spain), so the name *Homo calpicus* was at one time proposed for them. The modern name Gibraltar comes to us from Jabal Tariq, Arabic for Mount Tariq, named for Tariq ibn Zeyad, who along with General Musa ibn Nusayr led the Islamic conquest of Spain. The 1998 sesquicentennial of the cranium's discovery was commemorated with a paleontology congress held in Gibraltar and christened Calpe '98. To tell the truth, the cranium from Gibraltar was not the first Neanderthal fossil discovery. In 1829 or 1830, a two or two and a half-year-old child of the same human group had been found in Engis Cave, Belgium. Nevertheless, the discovery of the cranium at Gibraltar preceded the discovery of a skullcap and skeleton parts in the Feldhofer Cave in the Neander River Valley of Germany by eight years. Neandertal, or Neanderthal in an older German spelling, means "Valley of the Neander," and that is the name that stuck, because the earlier discoveries at Engis and Gibraltar had simply not attracted very much attention. It was the fossil skeleton discovered at Feldhofer that unleashed the controversy over the meaning of this species to our understanding of human evolution. As we shall see, it is a controversy that has continued up to our own day.

The scientific name *Homo neanderthalensis* was originated by William King, first used by him in a report read at the 1863 meeting of the British Association for the Advancement of Science. The existence of the Gibraltar cranium, which had in the meantime been sent to London, was recorded by the zoologist George Busk *sixteen years after its discovery*! Busk and the paleontologist Hugh Falconer presented their findings to the

BAAS at its September 1864 meeting in Bath. Falconer suggested to Busk that the Gibraltar fossil be designated *Homo var. calpicus*, a scientific name that never became official. In any case, the fossils from Gibraltar were of the same species that had already been named *Homo neanderthalensis* by King. The written version of King's report was published in January 1864.

The Neanderthals are without a doubt the best understood extinct human species. Erik Trinkaus has been able to study the age of death of no less than 206 individuals. On the basis of strong and extensive skeletal evidence, we can confidently describe their physical characteristics. But despite the abundance of fossils, we still have only limited and inadequate knowledge of some crucial skeletal parts, such as the cranial base and the hip. Each of these structures is of great importance to different aspects of our human condition, speech and childbirth, respectively. With respect to the hip, we are interested in the difficulty of childbirth for women, but also in the very immature state of the human fetus at birth, since the narrowness of the birth canal through which we must pass limits the size to which we can grow before being born.

Our information about the Neanderthals' Middle Pleistocene ancestors, who lived from 780,000 to 127,000 years ago, is even more precarious, particularly with regard to the postcranial skeleton, i.e. the skeleton from the neck down. Fortunately we have the marvelous discoveries of the Sima de los Huesos, in the Sierra de Atapuerca, to fill the vacuum of fossil evidence. There is a parallel lack of information about the precursors of our own species. We know our present day morphology very well, but we have hardly any relevant postcranial skeleton remains from the Middle Pleistocene. We would need an African Sima de los Huesos in order to fill out the fossil record of our direct ancestors.

Getting back to the Neanderthals, let us begin with a description and interpretation of their morphology before trying to discuss their psychology. Some of the ways in which the Neanderthals differ from us physically are mere archaisms, primitive attributes that our ancestors shared but that we lost at some point in our more recent evolution. To this extent we are innovators, while the Neanderthals simply had not changed. But the exact opposite also occurred. With regard to some other characteristics, our species has been conservative while the Neanderthals developed innovations during their hundreds of thousands of years of isolated evolution in Europe.

The modern paleontologist works to distinguish between different types of traits. Up until a few years ago, this was not the case. The result was a lot of confusion when it came time to establish evolutionary relationships among fossils. For example, modern humans of 100,000 years ago, discovered in Israel and whom we shall discuss in due time, displayed certain primitive characteristics along with other, quite modern ones. Because of their archaisms, these people were associated with the Neanderthals, who throughout their evolutionary process retrained these particular archaic characteristics while they developed their own innovations in other parts of the skeleton. The proto-Cro-Magnons of Israel are archaic modern humans, but were alternatively believed to be either a variant or descendant of the Neanderthals, or a Neanderthal-Cro-Magnon hybrid. Today, however, we can see that their modern characteristics clearly situate them on our evolutionary line and their archaic characteristics do not connect them to the Neanderthals.

The most important lesson we can derive from cladistics, the methodology discovered by Willi Hennig that I mentioned in the first chapter of this book, is to distinguish between primitive characteristics and evolved, or derived, characteristics. The latter are the ones that count. Now we have an opportunity to put this lesson into practice. We will see that this is a very difficult assignment, and somewhat speculative in the absence of an adequate fossil record. But this problem will no doubt be resolved as new remains are unearthed. I am confident that with the discovery of more fossils in the coming years and their accurate analysis, thanks to cladistics, we will learn a tremendous amount about human evolution. That is the good news. The bad news is that the more we know about human evolution, the more we realize how many branches it has and how extraordinarily complex it is.

My somewhat long-winded digression about primitive and derived characteristics should also help to make another point very clear. The Neanderthals were not simply primitive versions of ourselves. They were not lesser humans with very limited mental faculties. They had many characteristics in common with us, by virtue of the long evolutionary history that we had shared up to the point where our two lines diverged. But the European branch was not stagnant after the time of that divergence. It continued to evolve and produced the Neanderthals, who developed their own distinctive characteristics, just as we did elsewhere. The Neanderthals were not living fossils. They did not belong to the past, and they were not anachronistic. In their particular epoch, they were just as "modern" as our ancestors, the Cro-Magnons, were. The two species were simply different. In fact, in one important aspect, cranial capacity, Neanderthal evolution paralleled ours. They ultimately developed a brain as large or larger than our own.

Neanderthal cranial capacity, also called endocranial volume or cephalic volume, is determined on the basis of fossil crania, and since many of these crania are incomplete and need to be reconstructed, there is certainly some margin of error. Nonetheless, the available evidence leads us to believe that the average Neanderthal cranial capacity was greater than that of modern humans. In any event, it was no less. The largest known Neanderthal cranium was found at Amud, an Israeli site. In fact, its endocranial volume is the largest of any in the fossil record at 1,750 cc. If it is true that the Neanderthals' brain was independently evolving to an ever larger size at the same time that our ancestors' was, we may find ourselves studying the most fascinating story of all: that of two species, each having separately achieved human intelligence, who then came into contact with each other. Or should we say they collided? But we should not get ahead of ourselves. In order to assert that the two species' intelligence developed on a parallel track we must first rule out two alternative explanations. One is that the Neanderthals and our forebears inherited their large brains from a common ancestor. The other is that the two evolutionary lines, ours and the Neanderthals', both developed such large brains, and perhaps other characteristics in common, because we exchanged genes.

We are not yet able to provide a reliable figure for *Homo antecessor*'s cephalic volume. As I have said, it seems to have been more than 1,000 cc. We just don't know how much more. But there are a few fossil crania of what would have been the Neanderthals' ancestors in the European Middle Pleistocene. They may help us in our research. One of them, the Steinheim cranium from Germany, is quite deformed, making it difficult to determine its cranial capacity definitively. I would say that it just exceeds 1,000 cc, though it could be 1,100 cc. Nor do we know its age very accurately. It may be between 300,000 and 420,000 years old, although I think that the first figure seems closer to the truth. Another Middle Pleistocene cranium that was found in Europe is from Petralona, Greece. It is not possible to determine its age, but it has a capacity of 1,230 cc. Fortunately, Sima de los Huesos has provided us with three crania whose capacities can be determined with a very small margin of error. Cranium 4 has a capacity of 1,390 cc; Cranium 5 has a capacity of 1,125 cc; and Cranium 6 has a capacity of 1,220 cc. They are each about 350,000 to 400,000 years old, perhaps a bit more or a bit less.

What was happening meanwhile in the African Middle Pleistocene? In 1921 a very complete cranium was found at a place called Broken Hill, or Kabwe, in Zambia. Its cranial capacity is 1,285 cc. We do know its exact age for a change, but it does not seem to be one of the oldest of the African Middle Pleistocene. It may have been contemporary with the fossils recovered in Burgos from Sima de los Huesos. Fossil KNM-ER 3834, found on the eastern shore of Lake Turkana in Kenya, is comparable in age to the Broken Hill find, and its cranial capacity of about 1,400 cc considerably exceeds the latter's. Other Middle Pleistocene fossils found in Africa whose cephalic volume can be estimated include a 1,280 cc cranium from Florisbad, South Africa and one from Ndutu, Tanzania, measuring nearly 1,100 cc. There is also a fragmentary cranium from Sale, Morocco, that seems to have a lesser volume, perhaps 1,000 cc or less. While the Sale and Ndutu crania could be about 400,000 years old, older than the one from Broken Hill, the Florisbad fossil seems to be a little more modern, about 250,000 years old.

There is also a series of African fossil crania between 100,000 and 200,000 years old whose cephalic volume can be determined. They are not yet morphologically modern, but were evolving in that direction, and they can certainly be called premodern. They are the crania of Eliye Springs, on the western shore of Lake Turkana, Kenya; Omo Kibish 2, Ethiopia; Laetoli 18, Tanzania; and Jebel Irhoud 1 and Jebel Irhoud 2, Morocco. Their endocranial volumes range from 1,300 to 1,430 cc.

Based on these figures, I believe that we can deduce an increase in cephalic volume in both Africa and Europe during the period from 127,000 to 780,000 years ago known as the Middle Pleistocene. Some 300,000 years ago, some human crania on both continents reached volumes of about 1,400 cc, as in the case of Cranium 4 at Sima de los Huesos and of fossil KNM-ER 3834 from the eastern shore of Lake Turkana. The average among the respective European and the African populations was probably still under that of the Neanderthals or that of modern humans, since cephalic volume continued to increase during the last part of the Middle Pleistocene and the first part of the Upper Pleistocene, less than 127,000 years ago.

It is interesting to point out that the oldest European and African fossils from the Middle Pleistocene are associated with Acheulean industry, technical Mode II. But the first signs of another stone flaking technique, technical Mode III, appeared in Europe and Africa 250,000 years ago if not somewhat more. This was the beginning of the Middle Paleolithic. The Middle Paleolithic is characterized by this technical mode, the Levallois method, which involved the meticulous preparation of a stone core for the subsequent production of desired instruments in the form of flakes to be chipped off it. Thus it was a process with two clearly differentiated stages.

We can picture *Homo ergaster* chipping at a small stone and imagining the appearance of the biface he or she was intent on creating. In other words, inside our heads we have a mental representation of a *Homo ergaster* with a mental representation of a biface inside his or her head. In the visual language of cartoons, the biface would be represented inside a bubble, a large oval connected to the head of a prehistoric human by a chain of small circles. In any case, the *Homo ergaster* in question "sees" the result of his or her work before performing it, not with his or her two physical eyes, but with an "inner eye" or "third eye." We can say that he or she "transfers" that mental image to the stone or transforms the stone to match the image.

The practitioner of Levallois industry had to imagine both the prepared core, the first goal in the chain of operations, and then each instrument that would later be produced with one additional blow. There was a new complexity in this process that seems to reveal a greater capacity for planning than the previous technologies. And as we know, planning is a very human characteristic.

The first manifestations of technical Mode III coincided in time and space with the first human brains to reach a volume of 1,400 cc. We could posit a cause-and-effect relationship in this correspondence. But although such a hypothesis is tempting, it would be difficult or impossible to verify. What the archeological record does demonstrate is a certain technological continuity within Europe, Africa, and Asia as far east as the Ganges, evidence for cultural communication among the peoples of these continents. Nonetheless, regional variants of the technology are evident within the Middle Paleolithic, a pattern that is compatible with a model of geographical diversity in biological and cultural evolution and a certain permeability of boundaries, allowing the technologies developed by particular populations to influence others through contact. Mousterian technology was developed in Europe, western Asia, and North Africa, particularly in what is today Libya and Egypt. Other variants of Middle Paleolithic, or Mode III, are found in India and the rest of Africa.

At this point I would like to bring up a term minted by Pierre Teilhard de Chardin, the *noosphere*. Vladimir Vernadsky had introduced the term *biosphere* to refer to the totality of life as a multitude of independent yet interrelated beings in a closely woven mesh that forms a film around the Earth. Teilhard de Chardin believed that the human species forms its own film around the planet, a community of intelligent living creatures thinking in unison or connected by their minds, the outermost layer, in fact, of the biosphere. Teilhard de Chardin died in 1955, so he will never know how close his dream-metaphor came to fruition in the material reality that we know as the Internet. In the context of our discussion, the spread of Mode III Levallois-Mousterian technology throughout Europe, Africa, and part of Asia, following the similar spread of Mode II or Acheulean, could reflect the real existence of a noosphere. It may have been a very loose-knit web and it may have been circumscribed to only a part of the planet. Still, this may be evidence that cultural or at least technological ties reached beyond biological boundaries to encompass a broad area containing the African and Eurasian populations. And notice that the *Homo erectus* population beyond the Ganges in East Asia utilized an invariable Mode I technology throughout this period and does not seem to have been integrated into either of these two noospheres.

But there is another possible explanation for the archeological and paleontological evidence. Perhaps the European and African populations, as well as the Asian's geographically closest to them, maintained a biological connection all the while. They may have been generally isolated from each other, but still have been exchanging genes across their borders, the genes that produced the growth in brain size, for example. This is the multiregional model of evolution, whose principal proponents today, albeit with some shades of difference, are Milford Wolpoff, Fred Smith and David Frayer in the United States; Wu Xinzhi in China, and Alan Thorne in Australia. Their model entails the East Asian *Homo erectus* populations as well, although the increase in their cranial capacity was not nearly as marked. The most modern remains in the Javan fossil record are a series of fourteen cal-varia, or skullcaps, and two tibias found within the alluvial terrace of the Solo River in Ngandong. Their age is subject to debate. Although Carl Swisher's team asserts that they are between 27,000 and 54,000 years old, some geochronologists, including Christophe Falguères in France, say they are 200,000 years old or more. Two other Javan skullcaps that seem to be more or less contemporary with those from Ngandong have been found at Sambungmacan and Ngawi. The cranial capacity of six of the Ngandong skullcaps can be calculated or estimated. They vary from 1,013 to 1,251 cc. The capacity of the Ngawi skullcap is less than 1,000 cc and the one from Sambungmacan is about 1,200 cc.

There is a pair of 150,000 to 300,000-year-old fossils in China that differ from the classical *Homo erectus*. We cannot be any more specific about their age. One is a low-volume cranium from Dali whose capacity is about 1,100 cc. In some respects though, it resembles modern humans, in its facial morphology for example. We have a fragmentary skeleton found at Jinniu Shan that has barely been described yet, but that may provide interesting information in the future. Its estimated cranial capacity is greater, about 1,260 cc. These fossils could be taken to be migrants from Africa who displaced a local population of *Homo erectus*, but the multiregionalists see the possibility that they evolved from *Homo erectus* through the influence of African genes, that is by the mixing of genes as opposed to the complete replacement of one population by another. If it is determined that the Dali and Jinniu Shan fossils are contemporary with the youngest Zhoukoudian *Homo erectus*, we will be contemplating the coexistence in China of two distinct human types, perhaps even two species—one, *Homo erectus*, of local origin, and the second, represented by the Dali and Jinniu Shan fossils, having migrated from elsewhere. On the other hand, it could be determined that the latter fossils are younger. In that case, the multiregionalist hypothesis that *Homo erectus* evolved into a new human type with the help of outside genetic material will be more defensible.

China clearly presents a complex problem. Though it is a very large area, it has thus far produced few fossils and little conclusive evidence. We have only limited knowledge about what happened there between the time of the last *Homo erectus* of Zhoukoudian and the arrival of completely modern humans. The Chinese landmass may have much to tell us before long, but in the meantime I think the best strategy for this book will be to stick to what we know best. The species that serve best for examining an evolutionary puzzle like the one we face are those that can be examined independently of evolutionary entanglements. These species either reached extinction without producing a descendant species, or they currently exist, not yet having produced one. As it happens, the two human types that we know best fit into these categories. They are the Neanderthals and ourselves. I will begin by describing the former.

How to Recognize a Neanderthal on the New York Subway

It has sometimes been said that the Neanderthals resembled us so much, that dressed in modern clothing they could pass unrecognized on a New York subway. Carleton Coon famously included a drawing of a Neanderthal in a shirt and tie and fedora in his 1939 book. I know from firsthand experience that this is not true. I have never run into a Neanderthal on the Madrid metro, but I did participate in the reconstruction of a head based on a fossil from Sima de los Huesos for Javier Trueba's prizewinning documentary *Atapuerca: The Mystery of Human Evolution* (*Atapuerca: El misterio de la evolución humana*). I can assure you that it made a big impact on all of us when we masked one of our colleagues with a copy of the reproduction. When a primitive hominid is reconstructed, an Australopithecine for example, the creature that appears before your eyes is more familiar, less startling. It reminds you of a chimpanzee, even if it is a nonexistent bipedal chimpanzee species. But there is no existing equivalent to the Neanderthal, so similar to us, so human, but paradoxically so different. To come across a Neanderthal, even a reconstructed one, is a thrilling experience. It was no doubt even more astonishing to our ancestors, who met them in the flesh.

What was different about the Neanderthals? What did the Cro-Magnons think when they saw them for the first time? To begin with, the Neanderthals were very light-skinned and the Cro-Magnons were less so. How do we know the skin color of fossil humans? It is easier than you might think. Human populations that live near the equator get so much sun that it is very dangerous for light-skinned individuals. Fatal skin cancers can result, as indeed they do among light-skinned people today who expose themselves to prolonged and intense solar radiation. The populations native to the tropics defend themselves biologically, though, by producing a pigment called melanin in the bottom layer of the epidermis. Both Africans, who are called "black" for this reason, and Australian aborigines, who are not African at all, have very dark skin. In 1936 the Harvard anthropologist Earnest Hooten wrote that some strange and twisted psychology must motivate members of "the white races" to curl their hair and inconvenience themselves for the sake of a suntan, all the while interpreting naturally curly hair and pigmented skin as signs of racial inferiority.

Under the epidermis is another layer of skin, the dermis, which is where Vitamin D_3 is produced. Vitamin D_3 is essential for bone formation, but in order for the vitamin to be synthesized, the dermal cells must receive a sufficient amount of ultraviolet radiation. Despite their dark skin, Africans and Australians are exposed to so much radiation that they receive more than enough for this purpose. But Africans in the diaspora who live further from the equator and do not take supplements may suffer a Vitamin D_3 deficiency, with consequent complications of bone development and the possibility of rickets, which often poses a grave danger to women in childbirth. Far from the equator, people with light skin have the advantage, because they are better able to absorb and exploit the limited ultraviolet radiation available. So the Neanderthals, having evolved in Europe at medium and high latitudes, would have been lighter-skinned than the Cro-Magnons, who developed in Africa, at least until these new Europeans adapted to local circumstances, a process that would have taken some thousands of years. I have read only one novel about prehistoric humans written by a professional paleontologist. *Dance of the Tiger*, by Björn Kurtén, is a wonderful book. In it, the Neanderthals are called "the Whites" and the modern humans are called "the Blacks."

The Neanderthals had a well-developed supraorbital torus, a bony ridge above the orbits, or eye sockets, the cavities on the facial skeleton that house and protect the eyeballs. The supraorbital torus, or brow ridge, was not at all prominent on *Homo habilis*, but thickened and became conspicuous beginning with *Homo ergaster*. In fact, it is lacking in our species alone, although some *Homo sapiens* had still not completely lost it as recently as 100,000 years ago. But the more pronounced Neanderthal supraorbital torus was a truly distinguishing feature, forming a double arch above the orbits with no separation in the space between the eyes and above the nose. In other human species with supraorbital tori, the curvature above the orbits was less regular, and there was a central notch between the two arches.

Above the supraorbital torus, the Neanderthal forehead was low and sloping compared to ours. We don't know what function the supraorbital torus fulfilled, but there are several possibilities. For example, it may have protected the eyes from downward blows or absorbed the mechanical tension generated in the facial skeleton by chewing. The stress created by chewing is transmitted upward from the teeth and might have literally detached the face from the rest of the cranium at the forehead had it not been for the structural reinforcement provided by the torus. Unlike the foreheads of all other hominids, ours is a vertical continuation of the face on the same plane. This may enable it to dissipate the mechanical tension produced by our chewing as the supraorbital torus did for the Neanderthals'.

People often say that the Neanderthals' brow ridge and low sloping forehead made them look archaic and rough, while the absence of the supraorbital ridge and the high forehead of our ancestors the Cro-Magnons was more graceful. But Björn Kurtén sees the contrast from the Neanderthal point of view. That bony visor over their eyes provided them with a proud and fierce profile, like the daunting visage of the eagle. Modern humans though, with their lack of a brow ridge, their high foreheads, and their smaller faces, would have reminded the Neanderthals of their own children. In fact, both a prominent forehead and a small, minimally protruding face are universal features of immaturity among mammals. They elicit feelings of protectiveness and tenderness that inhibit adult aggressivity toward their offspring. In fact, these features also distinguish females from males of the same species, with the same emotional and behavioral implications. In cartoons, dolls, and stuffed animals, these childlike and/or feminine features are used to communicate likability. And in representations of adult females, they are often exaggerated along with the breasts and hips. It seems that these mechanisms are genetically programmed in all mammals, and we are no exception. If this is so, then the Cro-Magnons must have looked very cute to the Neanderthals! They may have discovered later, to their dismay, what kind of people they were dealing with, and as sweet as the Cro-Magnons may have looked, what kind of behavior they could expect.

The Cro-Magnon neurocranium or calvarium, the bony braincase, tended toward the spherical, as does ours. This is another trait generally associated with the young and with females. The Neanderthal braincase though, was very long, with the occipital bone projecting to the rear at the back of the skull. The Neanderthal neurocranium was characterized by a number of other innovations, which I won't detail, that are unknown in any other fossil. I don't know if there were any bald Neanderthals, but there probably were, because all mammals tend to lose hair with age. In any case, these cranial details would not be particularly striking under a head of hair. They don't elude the eyes of the paleontologist, though, and they tell us that the Neanderthals were quite singular and must have evolved separately from the other humans of their time.

The Neanderthal facial skeleton was morphologically unique and quite remarkable. The nasal aperture, the skeletal opening to the nasal cavity, was situated much further forward from the sides of the face than ours is. The nasal bones, which form the roof of the nasal cavity, were projected so far forward that they were nearly horizontal. The maxilla and malar bones together formed flat surfaces to the left and right of the nasal aperture, arranged at an angle that gave the face a wedgelike appearance. Cut on a horizontal plane, the cross section of a Neanderthal face almost suggests the triangular profile of a jet fighter with its wings sweeping sharply back.

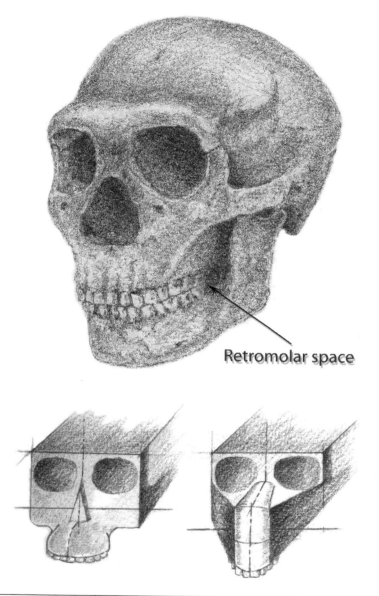

Retromolar space

Figure 10: The Neanderthal cranium. The lower diagrams compare Neanderthal morphology on the right with a hypothetical less-specialized human ancestor on the left. (Inspired by Rak, 1986)

Alan Thorne, whom I mentioned before, once spoke to me about Neanderthal physiognomy and said they had a "high-speed face." It looks aerodynamic, as though it were adapted for efficiency in forward motion. Needless to say, the Neanderthal face did not evolve for increased velocity, so we have been seeking other explanations. The biomechanical and climatic hypotheses are the most reasonable. According to the former hypothesis, the wedgelike facial morphology would have transmitted stress generated in the forward portion of the face off to the sides to be dissipated. This stress would have been generated by intensive use of the incisors, which have always been found to be very worn down among Neanderthals, even at young ages. It is not known why they would have used their front teeth so much, but it is supposed that they did so for paramasticatory or nonmasticatory purposes, for example to clamp objects as in a vise or to toss things backwards. This would have been as if they used their teeth as a "third hand."

The climatic hypothesis proposes that the Neanderthal face represents an adaptation to the extreme, almost polar cold of their environment. The enormous nasal cavity would have functioned as a radiator to humidify and warm the frozen air before it entered the lungs. Above and to the sides of the nose, the maxillary and frontal sinuses were extraordinarily well developed, so the face would also have been a kind of large hollow mask that insulated the brain from the bitter cold outside air. The image of the aerodynamic face seems to suggest that the lateral portions of the face migrated back to form the typical wedge shape that we talked about. In fact, this is precisely the evolutionary model proposed by the American Erik Trinkaus. The peripheral facial structure that would have receded was the part that supported the principal chewing muscles, the masseter and the temporalis, while the central portion containing the teeth would have stayed in place. Behind the third and last molar on the mandible, the

Neanderthals had a large gap called the retromolar space. According to the climatic hypothesis of facial adaptation, the retromolar space is the result of the separation between the part of the face used for mastication, the stable part, and the musculature, which receded.

Another researcher though, the Israeli Yoel Rak, does not believe that the periphery of the Neanderthal face receded with respect to the center. He believes that their facial morphology evolved for the biomechanical reasons offered above and that the retromolar space can be explained otherwise. While their molars became smaller, he says, the size of the Neanderthal mandible remained constant, leaving extra space behind the last molar.

The Cro-Magnons' chins must have attracted the interest of the Neanderthals. The chin is a strange outgrowth on the front of our jaw, and we are not certain of its purpose. The Neanderthals had no such structure. In this respect, it is we modern humans who present an innovation.

We may better understand the development of the Neanderthal face if we glance back at the human evolutionary record in Europe. The most complete Middle Pleistocene fossil cranium in Europe is Cranium 5 from Sima de los Huesos in Atapuerca. In fact, it is the most complete cranium in the entire human fossil record up to the Cro-Magnons. The other European Middle Pleistocene crania that retain their facial structure, some better than others, are one from Steinheim, Germany; one from Arago, France; and one from Petralona, Greece. Pardon the metaphor, but if the "high-speed face" of the Neanderthals indicates that they were cruising in overdrive, then their ancestors must have been chugging along in second or third gear at best. The sides of their faces were not as swept back. Cranium 5 and the other fossils from Sima de los Huesos must be ancestors of the Neanderthals, because they all share the same characteristic gap behind the wisdom tooth, the retromolar space. Also thanks to Cranium 5, we can settle the dispute about how the Neanderthal face was formed. It seems that there is something to say for each of the two hypotheses I outlined above. First, the molars got smaller and the face began to change. This is the stage represented by the Sima de los Huesos fossils. Later, there was a recession of the part of the mandible where the chewing muscles are inserted.

The 800,000-year-old fossils from Gran Dolina, Atapuerca provide us with the best evidence about the common ancestor of the Neanderthals and modern humans. It is curious that the most complete Gran Dolina fossil, that of an eleven-year-old boy, has a face that resembles modern humans much more than the Neanderthal face does, even though the Neanderthals came later in time. But our modern facial characteristics were always evident among prehistoric human children. Only with our own species does physiognomy remain constant throughout life. So yes, even adult Cro-Magnons must have looked very juvenile to the Neanderthals.

A detailed analysis of Neanderthal cranial features leads us to the conclusion that they were already present in two sets of 200,000-year-old remains found at Biache-Saint-Vaast, France. A German site at Ehringsdorf has also yielded cranial fragments with very Neanderthal characteristics, but their age is disputed. Some authors say that they are between 120,000 and 127,000 years old, while others say that they are older, about 250,000 years old. The latter may be closer to the truth.

The appearance of Neanderthal cranial features, subtle at first but increasing with time and categorical by the end of the Middle Pleistocene, is evidence for the European origin of the Neanderthals, a key question in evaluating the two models of human evolution presented above. I don't think that any African or Asian Middle Pleistocene fossil has these cranial features, even in incipient form, so I deduce that there was no exchange of genes between the Neanderthals' ancestors and other human populations of their time. They evolved in Europe in complete or nearly complete genetic isolation.

Elvis the Pelvis

The structure of the head and face was not the only feature that differentiated Neanderthals from Cro-Magnons. They would have been distinguishable from a distance by their body types and proportions. The Neanderthals were not more apelike, as you may think. The famous French paleoanthropologist Marcellin Boule, mentioned in the Prologue, erred in his analysis and reconstruction of the Chapelle-aux-Saints Neanderthal skeleton. He proposed a stooped, bent-kneed posture and a forward-slung neck, altogether less than erect. But Boule did not recognize the pathological nature of these postures. They were the result of disease, not of species morphology.

One way to organize the comparison of human body types is to superimpose hypothetical cylindrical templates upon them. The weight, volume, and surface area of a solid cylinder depends on its height and diameter. The relationship between surface area and volume is very important to the thermal regulation of mammals as animals who tend to maintain a constant body temperature. Two cylinders of the same diameter will necessarily have the same relationship between their outside surface area and volume, no matter what their height. But the greater the diameter, the lower the ratio between the outside surface area (the numerator) and the volume (the denominator). A proportionally small body surface is functionally advantageous for cold-weather animals, since it reduces the potential for heat-loss through the skin. In hot and dry climates, exactly the opposite is the case, and the ideal body type is represented by a narrower hypothetical cylinder. These differences can be significant among people. Let us imagine two people with the same volume and body weight, say 155 pounds. Their respective skin surface areas will be very different depending on their body shape. If one person is tall and thin and the other is short and stout, the former could have a skin surface 1.7 times as great as the latter. The former would dissipate a lot of calories in a cold climate, but for the latter it would be difficult to regulate his or her body temperature in a hot and dry climate, perspire as he or she might.

One way to calculate the diameter of an individual's hypothetical body cylinder is to measure the maximum width of his or her hips. The human pelvis is something like a ring formed by two coxal bones on the sides and front, and the sacrum, which completes the pelvic girdle in the back. The maximum diameter of the pelvis is a transverse measurement between the upper edges of the coxal bones, a dimension known in anthropology as the biiliac breadth. This diameter will also provide the approximate breadth of the entire trunk. The height of the body cylinder corresponds to the stature of the individual.

Unfortunately, there are not many fossil skeletons available that are complete enough to be measured for both pelvic width and body height. In fact, there are very few, so usually we have to estimate these dimensions and allow for a considerable margin of error. Again, we have been lucky enough to have found an abundance of pelvic material at Sima de los Huesos in Atapuerca. The most complete pelvis is from a male that we call Elvis.

I studied the entire postcranial skeletons of the human fossils from Atapuerca with José Miguel Carretero and Carlos Lorenzo, but I have to confess that from the neck down, my favorite part

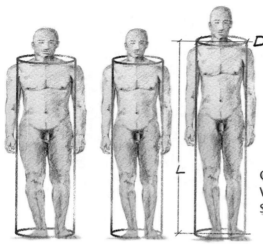

Outside surface area: π**D L**
Volume: $\pi/4$ x **D²L**
Surface / Volume: **4/D**

Figure 11: Body cylinders. Relative body surface decreases as the diameter of the body cylinder (D) increases, in accordance with Bergmann's biogeographical rule that cold-weather populations of a species tend to be heavier than its warm-weather populations, thus reducing the loss of body heat. (Inspired by Ruff and Walker, 1993)

of the skeleton is the pelvis. I have a passion for it. I did my doctoral thesis on this structure some years ago, before we had any of the Atapuerca fossils. Luckily for me, there is abundant scientific justification for a special interest in the pelvis. It can help us to understand the evolution of human bipedal posture and of childbirth in relation to fetal head size. The fetal head must pass through this bony ring, so we study pelvic dimensions to learn about the extent of newborn development in the various hominid species. But that is not all. The pelvis can also tell us the sex of a skeleton and even its age at death. As we shall see, it is also the key factor in estimating the body weight of a hominid fossil. For all these reasons, together with the real scarcity of fossil pelvises, they are as important as crania to our understanding of human evolution. Today, I would say, they are even more important.

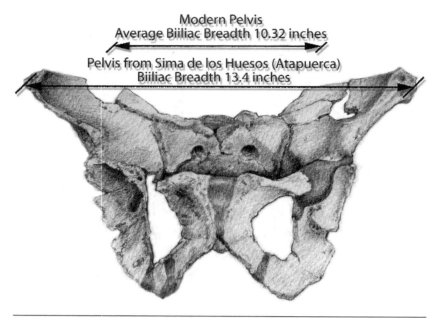

Figure 12: The Pelvis.

The American paleontologist Christopher Ruff has demon-strated a close association in our species between the weight of an individual (in good shape and without excess fat) and two other variables: the breadth of the hips and the height. Once this relationship was established, the formulas that had been worked out for living humans could be applied to fossils to determine their body weight. Ruff, Erik Trinkhaus, and Trenton Holliday have contributed to this endeavor.

The Neanderthals were shorter than us, with broader hips. One aspect of their physical type that attests to their adaptation to the cold is that their cubiti and radii, the bones of the forearm, and their tibiae, in the lower leg, were quite short. With their broad trunks, short forearms, and short lower legs, the Neanderthals would have struck the Cro-Magnons as very compact. This mor-phology seems to reflect Allen's law, which states that among mammals, humans included, the arms and legs, or members, become longer in warm-weather populations and shorter in cold-weather populations. The extreme of the former case would be found in an equatorial desert environment. Allen's law may seem to contradict what was said above, that the height of the body cylinder does not affect the relationship between outside surface and volume, but it does not because the human body corresponds not really to just one cylinder, but in a sense to the head and five cylinders, representing the trunk and the four extremities. Longer limbs add more to the body's surface area than to its weight, thus contributing to heat loss.

Using the above-mentioned relationships as a formula, the Neanderthals are calculated to have been quite heavy, heavier than modern humans of any race. The average individual would have weighed over 168 pounds, and many males would have exceeded 175 pounds. In proportion to their body weight, the Neanderthal brain would not have been quite so large. In fact, according to Ruff and his colleagues, it would have been relatively smaller than ours.

I think the question of relative brain size merits some clarification. When related species of different body weights, mammals for example, are considered as a group, it is seen that body weight and brain weight do not increase proportionally. Body weight increases more rapidly. So the bigger the animal, the smaller the brain will be in proportion to the whole body. Conversely, a mouse's brain is relatively larger than an elephant's. The mathematical relationship between brain weight and overall body weight in a group of more or less closely related species is $y=ax^b$, where y is the brain weight, x is the body weight, and a and b are constant values. In the case of mammals, the values of a and b have been calculated by the primatologist Robert Martin at 11.2 and 0.76, respectively. According to Ruff and his collaborators, Neanderthals had a brain 4.8 times the size that a mammal of their bulk would be expected to have according to the formula, while our brain is 5.3 times the expected size. This proportion is known as the cephalic coefficient. As you can see, both human species were well endowed with gray matter, but we moderns do have a small advantage. As I mentioned in the first chapter, I am wary of the cephalic coefficient when it comes to comparing very closely related species, since there may be various reasons for changes in body weight, among them the need to regulate body temperature. Such changes in body weight could alter the value of the cephalic coefficient without implying any change in intelligence. So I prefer to examine the evolutionary trajectory of species, and that is what I will try do.

The pelvises recovered at Sima de los Huesos are really very wide. In addition, there are several femurs in the collection that allow us to estimate heights between five feet seven inches and five feet eleven inches. Given these heights and trunk diameters, the weight of the males would have been enormous, at least 198 pounds. I believe that their true weight would have been considerably greater, approaching 220 pounds, because the Atapuerca humans of 350,000 to 400,000 years ago probably had a much more developed musculature than ours, as did the Neanderthals who followed them. Their skeletal weight, which accounts for approximately fifteen percent of body weight, was also probably greater than modern humans'. Body fat must also be accounted for. The physical condition of a professional athlete may not be a good comparison with that of the prehistoric human, who would probably have stored energy reserves in the form of fat at times when food was abundant. A runner or a cyclist does not do so, because he or she knows that there will always be enough food available for his or her body to do its work. This allows him or her to eat only the food necessary to support the necessary musculature and to provide necessary fats. The same is true for a cheetah or lion, who must attain optimum running speed in order to overtake a gazelle or a zebra. When these speed-dependent predators can no longer reach top velocity, their end is near. A prey animal's days are likewise numbered when due to injury, disease, or age, it can no longer outrun predators. Human hunting success, though, does not depend on optimal speed, but on strength, resistance, strategy, and technology. So I think it is most accurate to imagine the humans at Sima de los Huesos as endowed with reserves of body fat in good times and very lean in times of scarcity.

In proportion to their enormous weight, the brain size of the Sima de los Huesos humans was smaller than ours or the Neanderthals. Judging by the largest and smallest cranial capacities in the collection, Cranium 4 at 1,390 cc and Cranium 6 at 1,125 cc, their cephalic coefficient was between 3.1 and 3.8. But to evaluate their intelligence as a function of their evolutionary trajectory, we need to compare them to their predecessors. Thus we must factor in the body characteristics of earlier hominids. The earliest one that we can talk about is Lucy, an *Australopithicus afarensis*, who had extremely wide hips for her very short stature. We don't have a *Homo habilis* pelvis yet, but we can suppose that their proportions were similar to those of the Australopithecines. We have a very complete *Homo ergaster* skeleton from Nariokotome (WT15000), but it is from a boy about ten years old, so we have to try to estimate what his adult dimensions would have been. It is possible that he would have grown quite tall, to about six feet one inch, although it has recently been said that his ultimate height might have been considerably less as a result of the diseased condition of his spine. Still, we have to wonder what his height might have been had he been healthy. In any case, the important thing is that according to the reconstruction by Christopher Ruff and Alan Walker, this individual had narrow hips and long members, like a tall narrow cylinder, a biotype similar to that of the region's present-day inhabitants. Of course this similarity is due to the process of adaptation to a similar hot and dry climate, not to any particularly close evolutionary relationship between the Africans of 1.5 million years ago and those of today. But if the reconstruction is accurate, then the humans who populated Europe would have become shorter and more burly over time, perhaps as an adaptation to the cold, until reaching the proportions found among the Neanderthals. The Cro-Magnons, who evolved in Africa, would later introduce the narrow-hipped, long-limbed biotype to Europe for a second time.

But I don't see it that way, at least in terms of the hip dimensions and the consequent trunk size. The reconstruction of the Nariokotome boy's pelvis is based on very incomplete and somewhat questionable remains, from which a hypothetical adult size had to be calculated. There is an isolated coxa that I mentioned in Chapter Two, a hipbone of the same species that is so similar to the Sima de los Huesos fossils that I think the adult morphology was essentially identical in the two species. Based on this interpretation, I reconstruct two million years of changes in body and brain size with the following scenario. The *Homo ergaster* were strongly built, but their brains were proportionally small. The humans who left Africa maintained their physical bulk, and the European branch (the Neanderthal line), and the African branch (the modern line), each evolved larger brains independently, both to a greater extent than did *Homo erectus* in East Asia. About 350,000 to 400,000 years ago, the cephalic coefficients in both Europe and Africa had values of between three and four.

The brain continued to expand in both evolutionary lines. The Neanderthals adapted to the cold with shorter extremities and perhaps lost some physical bulk with a somewhat narrowed trunk. This would have caused a slight drop in body weight, entailing a further increase in their cephalic coefficient to almost five. On the African line too, the body type changed. Their hips narrowed and their body weight fell even more than the Neanderthals' did, so that their cephalic coefficient ended up somewhat higher, a little over five, although their crania themselves were somewhat less capacious. Based on this data, I have two conclusions. On the one hand, the people who were contemporary with the Sima de los Huesos fossils, both in Europe and in Africa, had considerably smaller brains than either us or the Neanderthals. On the other hand, it cannot be argued that the people on our evolutionary line had more developed brains than the Neanderthals did. Based on the fossil evidence, we have no reason to believe that the Neanderthals were less intelligent than we are. It would be more accurate to say that we weigh less than they did.

The earliest skeletons of the modern type are those of two proto-Cro-Magnon burial sites in Israel. Jebel Qafzeh and Skuhl, a cave and a rock shelter respectively, are both believed to date from about 100,000 years ago. Nevertheless, there are less complete but older fossils between 100,000 and 150,000 years old in Africa, such as those from Klaises River Mouth in the far south of the continent. Paleontological evidence points to Africa as our species' place of origin. Recent data collected by molecular biologists from our diverse populations does not contradict that presumption. The two Israeli skeletons are of our biotype and have our narrow hips. The attenuated pelvis of modern humans is manifested at two levels. First, at the upper edge, where the biiliac breadth is measured, and also lower down, between the acetabula, the joints where the femoral heads articulate with the pelvis. Greater breadth at the hips is associated with greater trunk diameter, which is reduced in our species, undoubtedly as an adaptation to heat. Remember, our origin is presumably African. But as the upper pelvis grew narrower, so did the lower pelvis, reducing the distance between the two femoral heads. Since we are bipedal, this process provided a significant biomechanical advantage in long distance locomotion for the following reason: When we walk, the entire weight of the trunk and of whichever leg is in the air pivots on the femoral head that is being supported by the pelvis. As the hips narrowed over time, the center of gravity came ever closer to the joint, saving energy with every step.

At the same time, the bones, which had been very thick since the time of *Homo ergaster*, became considerably lighter in the first modern humans. The marrow cavities inside the shafts of long bones expanded, and the cranium walls thinned. The cranial tori, or ridges, also disappeared or were dramatically reduced. The Neanderthals had magnificent hands that were able to exert tremendous pressure. They also had strongly curved femurs and radii with very large joints for both their arms and their legs. The skeletons of the first modern humans had become much lighter as a result of all the changes that they had undergone. The changes had also been accompanied by reductions in muscle mass. The physical characteristics of the Neanderthals and other "archaic" hominids (That's what they used to be called. I use the adjective simply to contrast them with modern humans, our species.) are subject to various interpretations, but all agree that they reflect great physical strength, a necessity in view of their lifestyle.

One is tempted to think that the first modern humans had evolved a less robust skeleton as a consequence of a changed lifestyle less dependent on physical strength. But our ancestors' morphological evolution did not coincide with any technological change. The modern humans at Qafzeh and Skhul produced Mousterian tools like those produced by the European Neanderthals or the Neanderthals who inhabited the Middle East after these proto-Cro-Magnons, as we shall see.

The first humans to arrive in Europe were tall and thin by Neanderthal standards, but they were still strong compared to their descendants. The Paleolithic would see a progressive reduction in their size and robusticity, a process that would continue through the Mesolithic and Neolithic as well. The research of Vincenzo Formicola and Monica Giannecchini tells us that during the first part of the Upper Paleolithic, before the glacial apogee of some 18,000 years ago, the average height of Cro-Magnon males was about five feet nine inches and that of females was around five feet four inches. These heights are not very different from those of Westerners today. But in the last part of the Upper Paleolithic, 18,000 years ago to 10,000 years ago, average male and female height fell to five feet five inches and under five feet one inch respectively, and continued to fall to five feet four inches and a little over four feet eleven inches in the Mesolithic. This progressive diminution has reversed itself in the last century. Our children are taller and stronger than ever. Perhaps the genes bequeathed to us by our distant relatives in the first days of the Upper Paleolithic carry a hidden and heretofore unrealized potential for development, probably frustrated by poor nutrition and significant consanguinity. With better nutrition and increased mobility, that potential is coming to fruition among the most recent generations. We are becoming Cro-Magnons again.

PART TWO

Life in the Ice Age

CHAPTER FOUR

The Animated Forest

That emotional response, that attraction; we often succumb to the temptation to stop and listen for who knows what as we pass through its green-tinged light. They tell us that the soul of the untamed land has enveloped us and touched our own souls.

Wenceslao Fernández Florez, El bosque animado

A Primate among the Oaks

We are so accustomed to the fact that humans live in every part of the globe that it seems no more than trivial. We are a ubiquitous species, able to survive in the most varied climates and landscapes of every continent. But the zoological group we belong to, the primates or the apes, evolved in very specific environments and has never included an environmentally flexible species other than ours. Primates have lived in the forest for more than 65 million years, so historically we are closely linked to it. As a matter of fact, the characteristics that we and all primates share and that distinguish us from other animals are adaptations that allow us to move around in the treetops. No primate species other than ourselves has ever adapted to a completely treeless environment. They have simply not been prepared for such a dramatic change. To tell the truth, there are a few other existing primates who share our eccentricity, the gelada baboons for example, who live on the grasslands of the Ethiopian plateau, the hamadryas baboons of the parched rills of Ethiopia and Somalia, and to a lesser extent, the olive baboons and patas monkeys of the largely treeless East African savannahs.

Although the European continent is extensively forested, it is home to no primate species other than humans. Of course other primates did live here before the arrival of humans, when the climate was warmer and supported other types of vegetation. But only one monkey withstood the rigors of the European Quaternary, the so-called ice age, along with us. This was the barbary ape, also known as the macaque monkey. Apart from a population introduced into Gibraltar by humans, they are now extinct in Europe, and their natural habitat is restricted to North Africa.

Botanical biogeography, the study of the geographic distribution of plants, provides a means to divide vegetation on earth into a series of hierarchically organized units. The most inclusive category is the *Kingdom*, followed by the *Region*. There are six floral Kingdoms in the world. The geographical distribution of primates effectively coincides with two of them, the Paleotropical and the Neotropical. The former includes Madagascar and all of Sub-Saharan Africa save for its southernmost tip, which belongs to the South African, or Capensic Kingdom. Monkeys live there as well. The Paleotropical Kingdom also encompasses parts of Asia including the Indian subcontinent, comprising Pakistan, India, and Bangladesh; as well as Burma, and both continental and insular Southeast Asia, Thailand, Laos, Cambodia, and Vietnam in the former and Indonesia and the Philippines in the latter.

The Neotropical Kingdom covers all of Central and South America, apart from the tip of the southern cone, which is part of the Antarctic Kingdom. The Paleotropical and Neotropical Kingdoms are uniformly hot and mostly located between the Tropic of Cancer in the north and the Tropic of Capricorn in the south. The principal reason for the nearly total absence of monkeys and apes outside the tropics is seasonal temperature variation, which becomes more marked as one moves away from the equator. Primates cannot survive long periods without fruit, green leaves, tender stems and shoots, or insects for nourishment. Seasonal climatic variation is a result of the inclination of the earth's axis, something that has always existed, albeit with minor variations. But the cooling of the planet over the last several million years has resulted in more dramatic seasonal variation and is another important factor in the current geographical distribution of primates. In lands far from the equator, winters today are colder than those of the past.

To the north of the Paleotropical and Neotropical Kingdoms is the Boreal, or Holarctic Kingdom, which comprises North America, North Africa, all of Europe, and all of Asia outside of the Paleotropical. Apart from the macaques of North Africa, no primates live in the Holarctic Kingdom outside the East Asian Region, made up of Korea, Japan, and a part of China. This occurs despite the Kingdom's widely varied landscapes, that include arctic tundra, boreal forest, temperate forest, Mediterranean forest, desert, and steppe.

Finally, the Australian Kingdom comprises Australia and Tasmania, places that primates never migrated to.

Zoologists also divide the Earth's land masses into kingdoms and regions according to the geographical distribution of terrestrial vertebrate species. The botanical and zoological biogeographic kingdoms and regions generally coincide, since in essence they reflect very closely related animal and plant histories. Every species has an area of origin from which it has dispersed. For a species to inhabit a region other than its own area of origin, it must first migrate there, or its ancestors must have done so. Secondly, the environment must provide the conditions that the species requires in order to prosper. Over geologic time, the earth's land masses have changed position dramatically, coming together and separating from one another as a result of forces deep below the surface. The geographical distribution of organisms also tells us about the geohistory of the earth's crust, because fossils constitute evidence of the contact and separation of species.

Zoologists identify three biogeographical realms. The Neogaean Realm includes Central and South America. Since the Neogaean was an island-continent for millions of years, the fauna there are quite distinct. They would be even more so had the space between North and South America not been bridged by the Isthmus of Panama some three or 3.5 million years ago. As a result of this geological event, there was an exchange of fauna, and many South American animal species became extinct due to incursions by North American predators or competitors. Among the South American species that did not experience extinction were the Platirrhini, or New World monkeys. No one knows how they arrived in the hemisphere in the first place. Perhaps a few individuals drifted across the Atlantic on fallen trees after powerful tropical storms swept them down African rivers and into the sea.

The Holarctic is another zoogeographic realm which includes all of Eurasia, Africa, and North America. It is divided into the Nearctic Region in North America; the Palearctic Region in Europe, North Africa, and almost all of Eurasia; the Ethiopian Region in all of Africa apart from its Mediterranean coastal strip, with the addition of the Arabian Peninsula and Madagascar; and the Oriental Region, the tropical southern and eastern part of continental Asia, along with Indonesia and the Philippines. Primates inhabit the Ethiopian and Oriental Regions, but not the Nearctic or the Palearctic, with the exception of the barbary apes and Japanese macaques.

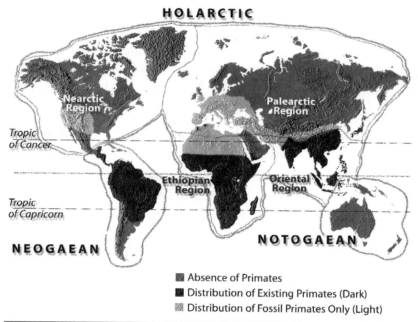

Figure 13: The Distribution of Existing and Fossil Primates: The boundaries of the faunal realms and those of the Holarctic regions are also illustrated. Today's primates live primarily between the two tropics, but primate fossils have been found much farther north, in areas that previously had hot climates.

Australia, New Guinea, Tasmania, and a handful of Indonesian islands form the Notogaean Realm. Their fauna is very distinct as a result of their prolonged isolation. Apart from humans, very few primate species cross Wallace's Line, the boundary proposed by the great naturalist Alfred Russell Wallace between the Oriental Region and the Notogaean Realm.

I would like to propose a synthesis of the flora and fauna of Europe, dividing the continent into large-scale ecological units known as biomes, which correspond to general landscape descriptors. In the north, there is a treeless tundra. The most characteristic mammals to be found on the tundra are the reindeer, the musk ox, the polar bear, the variable hare, the arctic fox, and the lemming, a small rodent that experiences massive population explosions every three to four years. All of these animals have or have had a circumpolar presence, extending from the north of Eurasia to North America and Greenland. South of the tundra, the taiga, or boreal forest, consists primarily of conifers and forms a second ring around the pole. The elk, the moose, and the wolverine, a carnivore of the mustelidae family that I will discuss again later, are typical taiga-dwelling mammals. The animals of the tundra and the taiga have a circumpolar presence because North America and Eurasia are very close to each other at these high latitudes of the Arctic Circle, specifically at the Bering Strait between western Alaska and the easternmost tip of Siberia. We will return to the Bering Strait because of the important role it played in the ice age.

The rest of Europe is almost completely covered by temperate deciduous forest or Mediterranean evergreen forest. We don't need to describe the characteristic animal species of these forests, because both types can be found on the Iberian Peninsula, and we shall see about them shortly. From Eastern Europe to Mongolia, passing through Central Asia and China, there is a continuous steppe, a sea of grass whose most typical animal species are Przewalski's horse, the onager, which is another equid; the saiga antelope, the Mongolian gazelle and other gazelles; the steppe skunk, which is a mustelid, and a number of gerbil-like rodents and lagomorphs, like the pikas. The last wild Bactrian camels, the ones with two humps, live further south, where the steppe gives way to the Gobi desert.

We would have to conclude from our investigation of European plant and animal distribution that it does not provide a favorable habitat for primates, except in the case of our species, which has adapted to life on every continent. As we have seen, our group of primates, the hominids, are native to Africa, and our arrival in Europe was relatively recent. We spent our evolutionary infancy in tropical African rain forests, and we acquired human characteristics in more open environments, i.e. scrub forests and savannahs with scattered bushes and trees. This was our first home and for a long time our only one. When humans arrived in Europe we had to adapt to local conditions that were very different from those in our ancestral African homeland. In fact, since humans arrived in Europe, temperate periods like ours have alternated with glaciations, long periods of very intense cold that brought about dramatic changes in the inventory of plant and animal life. In the course of our evolution, we first had to transform ourselves, in Africa, from arboreal and exclusively forest-dwelling primates, and then those of us who migrated to Europe had to learn to survive outside the tropics. If we did not exist, there would be no primates in the oak groves, the pine woods, or the beech forests of Spain.

But to learn more about the environment that humans continued to evolve in on the Iberian Peninsula, let us begin with the most visible life-forms on the landscape, the plant communities.

A Four-Part Outline of Spanish Vegetation Today

Almost the entire area of the Iberian Peninsula is environmentally suited to forestation. It was almost completely covered with trees before humans utilized fire and ax to clear enormous areas for cultivation and pasture and to harvest wood products. This attack on the forest to make the land available for farming and livestock began in the Neolithic and has continued increasing in pace ever since, especially during the twentieth century. Before the process began, the various human species had hardly modified the vegetation or the landscape where they lived. They lived in small and dispersed groups, hunting for meat and gathering vegetable foods. But the natural harmony of that time is gone for good. Strabon, a Greek writer and a contemporary of Jesus, is usually credited with the description of the Iberian Peninsula as so completely covered with forest that a squirrel could cross it in its entirety by jumping from tree to tree. Although the attribution is false, there is no doubt that forest covered more of the peninsula at that time than it does today, even though the grain fields and pastures of the Hispanic peoples already occupied large areas that had been carved out of the forests.

Botanists say that if left alone, all of Spain and Portugal would naturally develop into some kind of forest; trees would be its dominant climax vegetation. Each forested area would resemble one of the several types extant, albeit in reduced and/or threatened conditions, on our peninsula. Only on our high mountain peaks is it so cold for much of the year that trees cannot live, as the soil is usually frozen. In their place we find crawling plants, alpine grasses, and arcticlike bogs that vaguely remind one of the northern tundra closer to the pole. As one approaches either environment, the tundra or a mountain peak, the forest peters out where the median temperature of the warmest month is below 50 degrees Fahrenheit. This is known as the tree line. It is generally reached at about 7,550 feet in the Pyrenees, 5,600 feet in the Cantabrian Range, and at about 6,550 feet in the Betic Range and the Iberian and Central mountain systems.

There are also dry plains on our peninsula where rain is so scarce that hardly any trees grow, and those that do are very scattered. This is the case in the dry lands of the southeast, especially in Alicante, Murcia, and Almería, and in the area of Cabo de Gata. It is not the heat, but the lack of water that inhibits tree growth. With irrigation, tropical products are grown quite successfully in these areas. There are also arid areas in the central Ebro depression, with the additional disadvantage of a more continentalized climate and its much colder winters. The area around Los Monegros is a good example. Unfortunately, the destructive activities of humans have only increased the barrenness of these lands where the forest was already struggling at best.

To help us classify the vegetation of old Hispania, we can divide the territory into two large floral regions. They are both within the Holarctic Kingdom, and both of them extend far beyond the Iberian Peninsula. The two regions are 1) the Euro-Siberian Region in the far north, including the coastal Basque country and the Cantabrian Range, Galicia, northern Portugal, and the Pyrenees; and 2) the Mediterranean Region, covering the rest of the peninsula. Due to its northern location and the rain-bearing influence of the Atlantic Ocean, the Euro-Siberian portion of Iberia is cooler and more humid than the Mediterranean zone, which is generally hot and dry. A large variety of broadleaf deciduous trees like beech, oak, birch, hazelnut, maple, elm, linden, service tree, etc. predominate in Euro-Siberian Iberia. All of these species lose their leaves in the autumn after taking maximum advantage of favorable summer conditions, which include mild temperatures and soil that always retains at least some moisture. These trees may thrive with dry "bodies," but they do require moisture for their "feet." The forests of the humid portion of Iberia proclaim the passing of the seasons as their leaves turn from green in the spring and summer, to brown as they wither in the autumn, and drop to the ground in the winter.

Above the broadleaf forests of the Pyrenees are large coniferous forests of black and white pine. But firs similarly coexist with the beeches at lower altitudes. The black pine can be found at higher elevations than any other tree on the peninsula, frequently surviving at altitudes of 7,500 feet or more above sea level. The mountain forests are superficially similar to the vast belt of conifers, the taigas, that surround the southern boundaries of the tundra throughout the cold boreal reaches of northern Eurasia and North America. I say that they are only superficially similar, because although they may look the same at first glance, they are not always made up of the same species.

The insignis pine, also known as the Monterrey pine, is a species native to California, but it is very widespread in our north, above all in the Basque country. In Guipúzcoa it covers forty-six percent of the forested area and no less than sixty-two percent in Vizcaya. These expanses of non-native pines, along with large areas of other conifers and of eucalyptus, an Australian import, should really be considered tree farms rather than forests. Their biodiversity is far less than that of any naturally ocurring forest, and apart from the maximization of short-term economic gain, the latter are much more valuable in every sense.

Apart from a few remnant white pine forests in León and Palencia, there are no native pines in the northern band of green comprising the Basque country and the Cantabrian Range west to Galicia. Some Galician forests of maritime pine, also known as pinaster pine or resin pine, may be native, but we cannot be sure because this species has been widely replanted, so much so that it now covers more area in Spain than any other pine.

The villagers of Zamakola, in the Dima municipality of Vizcaya, told the great prehistorian and ethnographer José

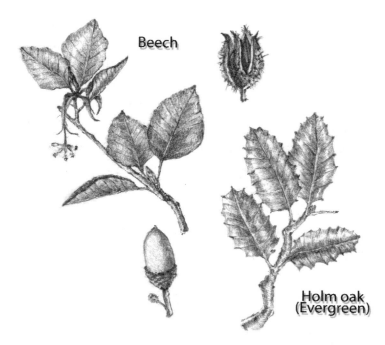

Beech

Holm oak
(Evergreen)

Euro-Siberian region (dark)
Mediterranean region (light)

CANTABRIAN SEA

CANTABRIAN RANGE

PYRENEES

IBERIAN SYSTEM

CENTRAL SYSTEM

ATLANTIC OCEAN

BETIC SYSTEMS

MEDITERRANEAN SEA

Figure 14: The two Iberias of the botanists.

Miguel de Barandiarán that the old pagan spirits in the village had been driven away by the bells of Christian chapels. There is a massive natural stone bridge in Dima called Jentilzubi, the Bridge of the Gentiles, that was believed to have been built by giant humans who inhabited the area before the Basques arrived. If I remember correctly, there is also a cave with two entrances called Balzoa. And there is a prehistoric site called Axlor, a shelter that was inhabited by Neanderthals, whose fossil remains have been found there. It was the same José Miguel de Barandiarán who excavated the site, the first such site I ever visited, when I was just twelve or thirteen years old. Having visited this place and others like it, so charged with history and legend, I know that it was the Monterrey pines and the eucalyptuses that drove out beings like the Galtxagorri, guardian angels so tiny that four of them could fit on a pincushion. The Galtxagorri had protected favored people, including the Lamias, sirens who spent their days brushing their hair alongside forest streams; Basajaun, the Lord of the Forest; Erensuge, the terrible man-eating serpent; and Mari, the goddess who inhabited Basque mountains and caves. There is neither magic nor mystery on the pine and eucalyptus plantations, where songbirds are never heard, and neither grass nor ferns carpet the ground. The fragile fairies of Basque mythology could not possibly find suitable abode in these monotonous expanses of identical trees where no mist swirls about overhanging beech, chestnut, or oak limbs.

There are fewer tree species in the forests of Mediterranean Iberia, but they are sometimes accompanied by impenetrable undergrowth, a much more dense and varied lower layer of brush than that found in the somber deciduous forests. The dominant trees, the Holm oak and the cork oak, have small broad leaves that are sclerotized, or hardened, with thick cuticles and sunken *stomata*, or pores. These adaptations help them to avoid the loss of moisture during the long dry summers they must endure. Larger-leaved deciduous trees would not survive such conditions. The Holm oak and the cork oak are evergreens; they are not leafless at any time of the year, and they can continue with their activities almost all the time, except when it is very cold. The cycle of the seasons is not as palpable in these evergreen forests as it is in the more humid areas of the peninsula.

Conditions for plants are especially difficult where the soil is loose and sandy or there are outcroppings of rock. Some places are exceptionally dry, and sometimes the climate is more continental and particularly variable, very cold in the winter and extremely arid in the summer. In all of these unfavorable environments, the Holm oaks give way to conifers like pines and prickly and Phoenician junipers. At this point I should mention the Aleppo pine and the stone, or umbrella, pine; both very drought- and heat-resistant and not very demanding when it comes to soil quality. But my favorite is the Spanish juniper, a marvelously austere and resistant conifer of the cypress family that can grow in the thinnest of topsoils and survive cold, heat, and drought. Managed juniper stands add a note of savage beauty to the most desolate regions of our barren inland plateaus.

However, the plant species of these two major regions of Iberian vegetation, one dry and the other humid, are not quite mutually exclusive. For one thing, there are Holm oaks on many parts of the Cantabrian coast, in drier locations, for example, or near the sea, which has a moderating influence on winter weather. There are also deciduous forests in some parts of the Mediterranean region where there is enough moisture year around. There are still beech woods, for example, on the Somosierra–Ayllón massif between the inland cities of Madrid, Segovia, and Guadalajara; and farther east in the Puertos de Beceite mountains between the coastal cities of Tarragona and Castellón.

There are conifers in the two Spains, as well. In the Iberian System, in the Cebollera Mountains of Soria and the Gúdar Mountains of Teruel, there are small black pine forests, and white pine is widespread in the Iberian, Central, and Betic Systems. In the Pyrenees and other mountains of the eastern half of the Iberian Peninsula, white pines cannot survive the summer drought. There are, however, stands of Spanish black pine, which is better adapted to the cold and dry conditions of medium and high altitudes in the Mediterranean region.

It would be difficult to find a better example of a Euro-Siberian enclave in the heart of the Mediterranean region than the famous stands of Spanish fir that survive in the Nermeja and Nieves Mountains of Málaga province and in the Pinar de Grazalema Mountains of Cádiz. This unusual fir species, like other closely related varieties around the Mediterranean, seeks refuge in mountains that capture precipitation by virtue of their altitude and their orientation with regard to weather systems. It goes without saying that their conservation is a top priority.

Moist river banks also provide a hospitable environment for advancing deciduous species in Mediterranean Iberia, including stands of poplar, ash, elm, and alder. Where streams are intermittent, the deciduous trees on their banks give way to copses of oleander and tamarisk.

Two Iberian broadleaf species marvelously exemplify an intermediate, perhaps "indecisive," ecological adaptation. These are the Pyrenean oak and the Gall oak, the latter closely resembling the Holm oak. They can be found in both the Euro-Siberian and the Mediterranean region. They are marcescent, which means that like the leaves of deciduous species, their leaves completely dry up in the fall. Many of them, however, do not fall from the tree until new leaves grow in the spring.

It may be more realistic to describe the distribution of flora in Iberia by delineating one zone of Atlantic influence, another of

Mediterranean influence, which would be the largest, and widespread intermediate zones in the interior, either Sub-Atlantic or Sub-Mediterranean. If the peninsula were a continuous plain, its vegetation would be more uniform and the transition between dry and humid territories would be more gradual. But the orographic complexity of the peninsula increases soil type, climate, and landscape diversity manyfold, as it always has. Spain can still boast of greater biodiversity than any other European Union country. In the beautiful words of Don Eduardo Hernández-Pacheco, "Rugged mountainous terrain is everywhere in Hispania; it is emblematic of our land: its forbidding precipices; its boulder-strewn highlands; its bold and craggy landscapes. From one end of the peninsula to the other; from the high Pyrenees to the southern Alpujarra; from green and rainy Galicia to the arid coasts of Almería; from the mountainous Catalan shore to the cliffs of Portugal's Atlantic coast, mountain chains and massifs intertwine without a breach."

This ecological diversity made it possible for hunters in the Iberian Quaternary to encounter animals adapted to rocky mountaintop and plateau environments, to forest and mountain meadows, and to more extensive grasslands, all within a limited territorial range. Unfortunately, this same wide variety of habitats within a small area makes it impossible for researchers to attribute any fossils found together at one site to a single natural environment, since human hunters and animal predators frequently consolidated their herbivorous prey at individual sites, regardless of its provenance. In Altamira, for example, we find fossils of the roe deer, a typical forest species, together with the remains of reindeer, which we place in the tundra and the taiga.

The Lost World

As we can see, the distribution of flora in today's Spain is governed by climatic conditions, the variables being temperature and precipitation. But it is not just the average annual temperature and the average annual precipitation that are determinant. The distribution of precipitation and freezing temperatures over the course of the year is also a factor. For example, the Mediterranean region's long dry summers significantly affect the prospects of various plant species that might otherwise thrive there. Due to the importance of climatic factors, the makeup of plant communities varies according to latitude and to altitude. Are they further north or south? Are they high in the mountains or at sea level? As we ascend a mountain system, we pass through a series of climates and plant communities somewhat comparable to what we would see if we were to travel from the Mediterranean to the North Pole. On the Iberian Peninsula this parallelism between mountaintop and more northerly vegetation is illustrated by the presence of species that have retreated and been confined to our higher elevations from the lowlands where they grew in colder times.

Soil type is also an important factor in the distribution of plant species. Although some plants are indifferent to the substrate, others are not. While gall oak and Spanish black pine have an affinity for the limestone found close to the surface across much of the peninsula, many other trees, the Pyrenean oak and the resin pine for example, do not thrive in limestone soil and prefer to avoid it. In any case, soil types today are substantially the same as they were a million years ago, so changes in the distribution of plant species since that time are due exclusively to climatic variation and just a few thousand years of human intervention.

In very general terms, the Earth's climate was hotter in the Miocene, between 25 million and five million years ago; and the Pliocene, between five million and 1.7 million years ago; than in the Quaternary, the most recent 1.7 million years. It was also wetter before the Quaternary. That is not to say that some regions and some periods were not more arid or more temperate than others. Remember, this is a broad generalization. But as you can imagine, the vegetation on the peninsula was different from today's. In a word, it was more "tropical," and yes, various kinds of monkeys and apes inhabited this land.

There were no temperate forests of oak, ash, hazelnut, or alder in the Iberian Miocene or Pliocene, but there were large forests containing many species that have no equivalents in the region today. On the Canary Islands, the Azores, and the island of Madeira, though, there are still some plant communities in areas unspoiled by humans that resemble some of the "tropical" forests that existed on the peninsula before the glaciations. These are the Laurisilva forests, cloud forests made up of leafy evergreens like the laurel. The leaf of the laurel is shiny and coriaceous, with a thick layer of cuticle. In order to survive as an ecosystem, the laurel forest requires moderate temperatures year round, and the constant atmospheric humidity provided by regular rain and mist. These conditions do not exist on the peninsula today, and they certainly did not exist during the coldest glacial pulsations. Under the circumstances, the laurel cannot form its own forest, but it still grows in a few especially favorable parts of the peninsula. Heavy mists form a suitable microclimate, for example, in the ravines south of Cádiz called *los conutos,* "the tubes." The Iberian madrone, or strawberry tree, is also a remnant of the Laurisilva forests of the Tertiary. Like the laurel, it has a close relative in the Canaries. Another species associated with the lost world of the Tertiary is a shrubby tree known as the English laurel or the cherry laurel, which grows in warm humid enclaves in both the Canaries and on the peninsula. Giant sequoias grew on our mountains in the Pliocene, but today they are found nowhere in Europe outside of parks and gardens.

The Ice Age on the Iberian Peninsula

The European climate must have been very cruel during the period of peak glaciation between 17,000 and 21,000 years ago. At one point sea level was about 400 feet lower than today. The Scandinavian icecap grew to a thickness of nearly two miles, and another one over Britain and Ireland grew to a thickness of one to 1.25 miles. Icebergs reached as far south as Lisbon. The average annual temperature on the Iberian Peninsula was about eighteen to twenty-two degrees Fahrenheit colder than it is today. To give yourself an idea of what these reduced temperatures mean, consider that in very general terms, average annual temperature drops one degree Fahrenheit for each seventy miles we travel to the north. This is called the latitudinal thermal gradient. It drops one degree Fahrenheit for each five and a half degrees of longitude that we travel eastward, away from the Atlantic. This is the longitudinal thermal gradient. And it drops one degree Fahrenheit for about every 275 feet we ascend a mountain, the vertical thermal gradient. To grossly simplify what this climate change meant, it was as if the Iberian Peninsula had been moved 1,250 miles to the north or had been lifted a mile above sea level.

If we could move the Iberian Peninsula 1,250 miles to the north, Madrid would be at the same latitude as northern Scotland. But this is a difficult comparison to make, because the highest peak in all of Britain is Ben Nevis, south of Loch Ness in Scotland, at 4,408 feet, while Spain has many higher mountains. We must also be mindful that climate is not controlled exclusively by factors as simple as latitude, longitude, and continentality, the distance from the sea. To illustrate, remember that Britain and Ireland are islands situated between fifty and sixty degrees north latitude, the same latitude as the Labrador Peninsula and part of Hudson Bay in Canada. But northwest Europe has a mild climate thanks to the Gulf Stream, which delivers warm water to European shores via its extension, the North Atlantic Current. The Canadian climate, on the other hand, is influenced by the cold waters of the Labrador Current, which carries water from near the Arctic Ocean. Ocean currents are so important to climate that some authors relate the emergence of the Isthmus of Panama between three and 3.5 million years ago to the beginning of the planetary cooling trend clearly detected in many regions some 2.8 million years ago. When North and South America were united by what is today the isthmus, the waters of the Atlantic and Pacific Oceans were isolated from each other. In the view of these climatologists, the resulting radical changes in the circulation of ocean waters led to the formation of large mantles of ice in northern lands.

It seems that at the time of the glacial maximum, the average annual temperature at altitudes of over 2,300 feet on the Iberian Peninsula was less than 37.5 degrees Fahrenheit. The tops of the major peninsular mountains were covered by perpetual snow. It is very difficult to know just how far down the mountainsides these perpetual snows reached, but since this does provide a rather intuitive way for people to gauge the impact of the glaciations, I can provide some rough figures. The large quantities of snow that accumulated above the altitude of 4,900 feet on the Montes de León and on the Picos de Europa did not melt at any time of the year. The snow above a similar altitude lasted year round in the Sierra de la Estrella at the western end of the Central Range. As we move east, the perennial snowline rose to 5,900 feet in the Sierra de Gredos and to about 6,550 feet in the Sierra de Guadarrama. The permanent snow line was higher in the Sierra Nevada, perhaps at an average of 7,900 feet, in the western Pyrenees at about 4,900 feet and in the eastern Pyrenees at about 6,900 feet. Within any east-west mountain chain on the peninsula, the year-round snow was lower in the west and higher in the east, since snowfall decreases in intensity and frequency as the influence of the Atlantic diminishes.

In some high mountain locations, the snow that is deposited in natural basins or in cirques, which are natural amphitheaters, accumulates and turns to ice. This is how a mountain glacier is formed. It may remain in the cirque or descend like a tongue of ice, retained by the walls of a valley. The glacier gradually reshapes the valley, the friction produced by moving ice wearing away at the valley floor and walls. The famous Ordesa Valley in the Pyrenees was once filled with an enormous glacial tongue that gave it its characteristic troughlike shape. Glaciers can descend more than a thousand feet below the perennial snowline on a mountain. On their way down they pull many stones loose, drag them along, and later deposit them at the bottom, the sides, and at the leading edge of the tongue, where the ice melts and releases liquid water. This frontal debris forms into ridges called moraines. The moraines and the shapes sculpted by the moving mass of ice are key to what we know about the advances and retreats of ancient glaciers.

During the coldest period of the last glaciation a small icecap even formed on a plateau in the Montes de León. Glacial tongues spread outward and downward from it. These tongues dug out Lake San Martín de Castañeda and Lake Sanabria in Zamora, at an altitude of about 3,300 feet, and they deposited terminal moraines at their greatest extensions. A small icecap also formed on a plateau of the Ancares massif.

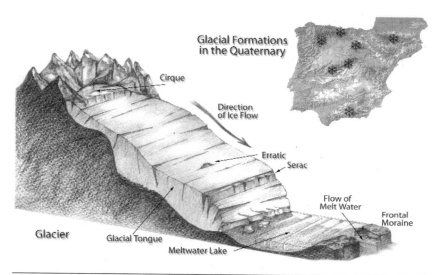

Figure 15: Glacial Formations.

There were many mountain glaciers of both the cirque and valley variety on that cold Iberian Peninsula. They developed in the Pyrenees, in the Central System, in the Sierra Nevada, in the mountains of Galicia and León, in the Cantabrian Range, and in the Iberian System. Above all, large valley glaciers, like those to be found today in the Alps, formed in the Pyrenees. Some of them reached lengths of twenty miles and thicknesses of more than a quarter of a mile. Glacial development was extensive in Gredos, despite its central location on the peninsula. Valley glaciers developed, and some of the highest peaks were even helmeted with ice. On the other hand, in the Sierra de Guadarrama and in the Iberian System, including the Moncayo massif and the La Demanda, Urbión, Neila, and Cebollera Ranges, glacial structures were small, little more than tongueless cirque glaciers, hanging glaciers, niche glaciers, or even smaller formations. The southernmost glaciers in Europe formed in our Sierra Nevada.

They included many cirque glaciers and some valley glaciers, like those at the headwaters of the Lanjarón and Genil Rivers.

In the Pyrenees, the Ancares, the Cantabrian Range, and the Sierra de la Estrella, some valley glaciers descended to below 3,300 feet above sea level. On the other hand, the glacial front remained above 4,600 feet in the Sierra de Gredos, above 4,900 feet in the Iberian System, and above 5,400 feet in both the Sierra de Guadarrama and the Sierra Nevada. Not a single live glacier remains on the peninsula today, with the exception of some very small and localized ice structures in the Pyrenees. Even they are retreating rapidly. The Marboré, Cilindro, and Monte Perdido glaciers are the best examples.

How many times have glaciers formed on the Iberian Peninsula in the course of the Quaternary? Almost certainly not with every one of the Northern Hemisphere's glaciations. Hugo Obermaier believed that he had found evidence of two glacial advances in the Sierra de Guadarrama of the Central System and on the Picos de Europa, which he attributed to the last two glaciations, the Riss and the Würm, following the pattern observed in the Alps. The Riss, the first of the two, produced the greatest glacial advance in the Spanish mountains.

Javier de Pedraza and other geomorphologists who have studied the glaciers of the Central System have identified only two significant pulsations. The glaciers there reached their maximum extension with the first pulsation and stabilized at a somewhat higher altitude with the second. All evidence leads these scientists to believe that the first of the two pulsations represents the coldest point of the most recent glaciation, after the disappearance of the Neanderthals, and the second represents the end of the Pleistocene. Juan Carlos Castañón and Manuel Frochoso have studied the glacial record of the Picos de Europa and have likewise concluded that the greatest advance was at the climax of the most recent glaciation.

So it is possible that neither the Neanderthals nor their ancestors ever encountered a glacier on the peninsula, apart from those in the Pyrenees. Their climatic environment was not as cold as that which the Cro-Magnons later had to endure. In the Sierra Nevada, researchers have identified some degraded and therefore questionable moraines that may have been produced in the penultimate glaciation, during the time of the Neanderthals' ancestors, but the widest range of cold-weather fauna on the peninsula seems to date from Cro-Magnon times. This supports the strong evidence that widespread Iberian glacierization was restricted to that period.

Each time that glaciers advanced southward into Europe during the Quaternary, the surface features of the continent experienced dramatic changes. Heavy mantles of ice covered much of its northern lands. The accumulation of water in the form of ice caused the sea level to drop dramatically. The English Channel was left dry, so that one could walk to England and continue on foot to Ireland, although only the southern portions of those two islands were free of the icecap. On a broad band to the south of the glacial front, the climate was what we call periglacial. Its outstanding characteristic is that the ground is permanently frozen dozens of feet deep, a phenomenon known as permafrost. Permafrost can reach a depth of up to 1,000 feet in Alaska and even more in Siberia.

Trees cannot sink their roots into frozen ground, so the tundra is covered with moss, lichens, and grasses instead. The air temperature does rise above freezing on summer days, thawing the ground to a depth of ten or twenty feet and creating a soggy uppermost layer. Large areas of standing water, unable to filter through the frozen and impermeable substrate, are trapped on the surface and form swampy lakes. During the glaciations, the ground conditions and vegetation at the foot of large and perennially snow-covered massifs were similar to those of the periglacial environment.

To the south of the tundra, the European continent was covered with immense coniferous forests like those of the taiga, the boreal forest. Further inland, large areas unaffected by the moderating influence of the sea were subject to the continental weather pattern of low precipitation and dramatically contrasting temperatures. This produced an additional treeless area with little ground cover, to which the wind carried and deposited enormous quantities of dust picked up from glacial deposits far away. This airborne debris produced the deep silt deposits called loess that provide fertile soil for today's large-scale grain production.

Finally, in some milder and more humid enclaves of southern Europe, deciduous forests of oak, beech, and other species held on, and Holm oak survived on the warmest parts of the Mediterranean coastline. Each awaited the opportunity of another climate change to expand once more at the expense of the tundra, the taiga, and the frigid steppes.

All of these climatic alternations had their effects on the Iberian Peninsula. Due to its latitude, the glaciations had less impact here than in more northerly lands, but the effects were considerable nevertheless. Thanks to studies of fossilized spores and pollens preserved in bogs, on lakebeds, and in caves, we are reasonably familiar with the evolution of vegetation on the peninsula from the period of the last glaciation to the present. At the coldest time of the last glaciation, from 17,000 to 21,000 years ago, the Cantabrian Range and the Montes de León were inhospitable to vegetation at altitudes above 3,250 feet. Few trees survived there, apart from some clumps of juniper and others like the creeping Phoenician juniper, mountain pines like the white pine and the black pine, and silver birch. These species hung primarily on south-facing hillsides and in deep valleys. The land was bare and frigid there, largely abandoned by animals and humans during the long winters, but large herbivorous mammals grazed there during the short summers. They were closely followed by four-legged and two-legged hunters. The situation was similar up to a thousand feet or so below the permanent snowline on the peninsula's other large mountain chains.

In addition to the highest mountain summits, broad swaths of the Spanish interior are at altitudes of well over 2,000 feet. The Iberian Plateau; the plateaus of Castile, La Mancha, and Alcarreña; and the Sierra Morena fall into this category. Cold steppes developed in these areas, with only a few scattered trees, much as on the mountainsides described above. The interior lowlands, on the other hand, were covered with thick coniferous forests. All in all, pines were the dominant tree species during this glacial epoch. Still, on the seaward slope of the Cantabrian Range, in other favorable microclimes along the Atlantic coast, and in Cataluña, there were enclaves of mixed deciduous forests that included oak, alder, hazelnut, ash, service tree, elm, maple, beech, and other trees. A narrow stretch of the eastern coast provided a haven for Mediterranean vegetation. In the last 10,000 years, from the end of the Pleistocene, during what is known as the Late Glacial, and above all during the Holocene, mixed deciduous forests and forests of Holm oak and deciduous oaks have expanded throughout almost the entire peninsula, relegating coniferous species to the least favorable environments, those that suffer from a lack of rain, poor soil, cold weather, or any combination of these disadvantages. The result of this process was that all Iberia became a forest.

Up to this point we have seen how climate change has affected the Iberian Peninsula, modifying the floral communities that its human inhabitants have known. But we have yet to discuss the animals that contributed to these ecosystems. Fossilized bones, the usual evidentiary resource in paleontology, provide just a portion of what we know about these animals. We also have the artistic representations that our ancestors left us, and another extraordinary kind of fossil, the frozen bodies of some animal species.

CHAPTER FIVE

The Reindeer Are Coming!

*At the crossing place Nahiktartorvik on the lower Kazan River,
there is a small rocky knoll which gives the place its name, 'the
lookout'. It is from there that the first glimpse can be obtained of
the caribou herds making their way northwards, and, when the
time is approaching, hunters often set out in sledges from the adjacent camps in order that they too may be ready to participate in
the joyful event. We happened to be at one of those settlements
when the sledges returned and the cries of 'the caribou are coming' were already ringing among the snow shelters as they first
came into sight.*

Kay Birket-Smith, Eskimos

The Mammoth That Came in from the Cold

On May 3, 1901, three travelers boarded a train in the Russian
city of St. Petersburg. They were headed for Irkutsk, on the
shores of Lake Baikal in Siberia. (The Czar's courier, Michael
Strogoff, had been sent there from Moscow in Jules Verne's well-
known novel named for his character. Russia was in the midst of
a violent Tartar rebellion led by Feofar Khan and the traitor Ivan
Ogareff, so the courier's journey was enormously perilous.) But

the historic adventure of our three passengers began in Irkutsk, rather than ending there. Their names were Otto Herz, a zoologist and the team leader, D. P. Sewastianoff, a geology student, and Eugen Pfizenmayer, a taxidermist. They were not secret agents, but researchers at the Russian Imperial Academy of Sciences in St. Petersburg. Their goal was to retrieve a frozen mammoth that had been found by a Lamut hunter next to the Beresovka River in the middle of the previous August and bring it back to the Academy. The Beresovka is a tributary of the Kolyma, feeding it from the southeast, above the Arctic Circle. The Russian Finance Minister had provided 16,300 rubles to the courageous Herz group for the expedition. The following summary of their story is based on a book published by Herz in 1902.

The scientists left Irkutsk on horseback and traveled 2,800 versts to Yakutsk, a town on the Lena River. One verst is equal to about 0.663 miles, so this stage of the journey was 1,856 miles long. They then traveled down the Lena by steamship as far as the mouth of the Aldan River, which they followed upstream, also by boat. They disembarked in Khara-Aldan on June 22 and resumed their journey by horse, this time to Verkhoyansk, another 622 miles, where they arrived on July 9. Another 1,425 miles by horse and boat brought them to Mysovaya on August 30. From there it was just 130 versts or about eighty-six miles to the location of the enormous animal, where they finally arrived on September 9. Otto Herz and his struggling party had brought twenty horses from Khara-Aldan, and had been accompanied by two Cossacks and three guides, one of whom was swallowed, horse and all, by the waters of a tributary of the Aldan River.

Once they found the mammoth they were confronted with two problems: how to preserve it, and how to transport it. Herz considered drying it in the open air or treating it with alum and salt.

Finally, though, he decided to bring it back to St. Petersburg as quickly as possible. If this could be accomplished before the advent of spring, then it would not thaw, and they would not have to concern themselves with its preservation. The mammoth was cut into pieces and loaded onto sledges to be pulled by horses and reindeer. In order to arrive in the capital before spring with their cargo intact, they would travel day and night.

The group set out on October 15, but conditions were even more dire on the return trip than they had been on the way out. Not only were they loaded down with over 3,600 pounds of frozen mammoth, but they were confronted with punishing Siberian autumn and winter weather. It was particularly difficult for them to cross the snow covered mountains that lay between them and Verkhoyansk in mid December, at temperatures hovering between -40 and -60 degrees Fahrenheit. They trudged along, helping the bedraggled reindeer who struggled to pull the sleds. The heroic expedition finally reached Irkutsk, where their precious cargo was loaded onto rail cars bound for its final destination, the Zoological Museum in St. Petersburg. It can still be seen there, reconstructed by taxidermists as though it had been dried intact. It was a grand occasion when the members of the expedition arrived by special train at the St. Petersburg station on February 12. Aside from their travels on train and steamship, they had covered almost 4,000 miles by sled and almost 2,000 miles on horseback, all within a period of ten months.

How did such an extraordinary fossil come to be? The frozen mammoth was not encased in a block of ice, as one might think. There are no giant blocks of ice containing mammoths in the frozen Arctic. We find preserved mammoths when they have been buried in the frozen Arctic earth, the permafrost, until human activity or the erosive action of a river reveals them. One day long ago, one of these gargantuan animals, a male, died in a gloomy, sunless riverbed, or somehow his body ended up there. His body then dried out in the freezing cold, undergoing natural mummification. The uppermost layer of permafrost thawed in the sun one short summer, and since the underlying frozen layers were impermeable, the resulting liquid water saturated the uppermost soil. The soil became unstable and eventually slid downstream. This process, known as *solifluxion*, is very common in periglacial environments. Several of these mudslides must have buried the Beresovka mammoth sufficiently to keep it frozen until our time.

The preservation of this and other mammoths was so complete that we have recovered not only bones, but more or less complete remains of their skin, hair, flesh, and even viscera. Thus we know, in many cases, what the animal ate shortly before death. The first frozen mammoth was found on the Bykovskiy Peninsula of the Lena River Delta in 1799. Another very famous find was a baby mammoth found intact in the Kirgilyah Valley in 1977. The main food of this six to eight-month-old male was mother's milk, but his teeth were already somewhat worn, indicating that he had also begun to graze. His empty stomach and lack of body fat lead us to believe that in his last days he must have been very hungry and may even have starved to death. Perhaps he got lost or his mother had died before him. As I write these lines, the news media are reporting the attempt by a Franco- Russian expedition to remove an intact and very well preserved mammoth from the Taimyr Peninsula of Siberia and transport it to Khatanga, site of a future ice cave museum, but with the arrival of the Siberian winter, the team has only been able to recover the head. It is awesome to think that these huge and hairy animals once inhabited almost the entire Iberian Peninsula!

The mammoths encountered by our Cro-Magnon ancestors in Iberia were *Mamuthus primigenius*. They were not as tall as today's African elephants, but they were still imposing and muscular pachyderms with large spiraled tusks. Their heads were somewhat pointed. The highest point on their backs was at the withers, descending sharply to the hindquarters. Naturally their ears were small compared to other Proboscideas, since large ears would have wastefully radiated a lot of body heat in their cold environment. These mammoths also had a considerable amount of body hair to protect them from the cold; they were woolly mammoths. The frozen specimens have light brown or yellowish skin, and they are frequently portrayed that way in illustrations. But we know from other dried and long preserved bodies, including human mummies, that dark skin pigment oxidizes and reddens with time, so it is probably better to imagine the mammoths as black. They certainly had long hair. We have found some hairs a yard long, and they were additionally covered with a fine down. Under their skin they had an insulating layer of fat one to two inches thick.

Legend has it that the expeditionary team at the Bersovka River site banqueted on mammoth meat. The reality is more prosaic. It seems that just one of them tasted a well-seasoned bit of the meat in the name of science. It did not stay down long. While the Beresovka mammoth was not really edible to even the most heroic researcher, there is another case of ancient meat keeping its flavor. In 1976 a 36,000-year-old bison was found buried in the Alaskan permafrost. It was nicknamed Blue Babe, for after its death, its skin turned blue as a reaction to the minerals present where it had long lain. This animal was a *Bison priscus*, the same species painted by ancient humans on the walls at Altamira and in many other caves.

Today there are two closely related species of bison, one in Europe, and one in North America. In fact it is possible to cross them and produce fertile hybrids. During the glacial periods, the sea level dropped considerably, continental shelves emerged, and dry land expanded its reach. The Bering Strait became a traversable land bridge between Asia and North America. Even today, the shortest point between the two shores is just about fifty-five miles. The Strait was part of a large intercontinental area exposed by the retreating waters. This area is known today as Beringia. It stretched from Siberia's Lena River to the Yukon in Canada. Blue Babe's ancestors migrated to North America by this route, as humans did more recently, some 13,000 years ago. Like the bison, the humans came from Asia. The North and South American Indians, or Amerindians, are closely related to the peoples of East Asia, like the Chinese, Koreans, Japanese, and Vietnamese. During the Holocene, the sea level rose again, forever separating the Eurasian and American bison populations. They, and the human populations as well, began to differentiate themselves.

But you may be wondering whether Blue Babe's meat was really edible. If we are to believe the paleontologist Björn Kurtén, who tasted a stew made from Blue Babe's meat, the answer is yes. He found that under the bluish skin of the bison, the meat was fresh and red, pleasantly flavored, if somewhat "earthy." But there is more. Kurtén and his colleagues were not the first to sink their teeth into Blue Babe's flesh. Some 36,000 years ago, a group of lions had killed the beast, leaving claw and fang marks on its body. Lions and woolly mammoths had also crossed into North America over the land bridge. The lions spread as far as Peru before dying out in the Americas. But Blue Babe's killers were unable to consume their prey because its body froze quickly in the intense cold, perhaps as night fell, and became so hard that it was abandoned nearly intact. Some time later, the bison was naturally buried in the permafrost and preserved for the ages. But not before a hungry lion broke off part of a carnassial molar that scientists have recovered from the skin of the frozen animal.

The Age of the Reindeer

The woolly mammoth is the best known representative of the ice ages. When the last glaciation ended, the mammoths disappeared with it, or almost did, as we shall see later. Another equally hairy large mammal was the woolly rhinoceros, *Coelodonta antiquitatis,* which also became extinct when the icy Pleistocene was no more. Although these two large herbivores did not find the habitat they needed to survive the climate change, their contemporaries, the reindeer and the musk ox, have found refuge in the Great North right up to our time. The former is found today in Eurasia, Greenland, and North America, and the latter only in Greenland and North America. Despite its name and appearance, the musk ox is not related to the cow, bull, or bison. It is not a bovine. Zoologists tell us that although male musk oxen may reach over nine hundred pounds, they are most closely related to the caprines, the group of animals like sheep, goats, Rocky Mountain goats, and chamois. The arctic fox, which turns completely white in the winter, is another refugee found in the far north. Its remains are sometimes associated with those of extinct herbivores in paleobiological digs. The North American reindeer, called caribou on that continent, migrate en masse twice a year. Like the prehistoric European hunters, Indians and Eskimos paid close attention to the movements of the great caribou herds. We suppose that the mammoths and woolly rhinoceri also left the frozen tundra and migrated to more favorable climes for the winter, returning to the soggy pastures of the north or to higher elevations in the summer.

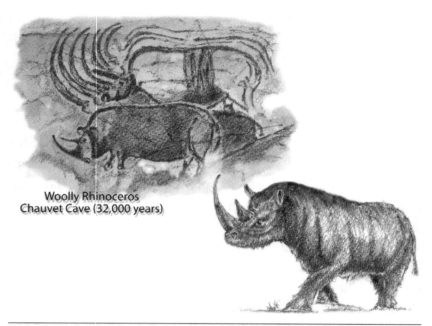

Woolly Rhinoceros
Chauvet Cave (32,000 years)

Figure 16: Reconstruction of the woolly rhinoceros and some paintings in Chauvet Cave. Fear-inspiring animals such as lions, woolly rhinoceri, and woolly mammoths abounded in the earliest cave paintings.

The maritime influence is a very important climatic factor. It not only moderates temperatures, but also provides the moisture necessary to produce rainfall. The farther inland one goes, the more continental is the climate, which means that it rains less, because air currents have dropped their loads of moisture before reaching that point. At the same time, temperature fluctuations are greater because they are not mitigated by the enormous mass of seawater. In interglacial periods like our own, the northern shores of Central Europe are influenced by the presence of the Baltic and North Seas. The former is practically a salt lake narrowly joined to the latter. One important reason for the harshness of the Central European climate during the glaciations was that the Baltic was completely frozen, while the North Sea froze to some extent and certainly retreated due to the lower sea level. This diminution of maritime influence or increased continentality, along with proximity to the large Scandinavian icecap and the smaller ones in Britain and Ireland, combined to produce an extremely continental climate in Central Europe. The northern tundra extended to the south, forming a continuous band with the cold dry steppes.

To tell the truth, the mammoths, the reindeer, the musk oxen, and the woolly rhinoceri roamed over both the tundra and the steppes, so much so that some authors call this group of species "the tundra-steppe mammoth fauna." Another species of the frozen steppe that also extended its range throughout much of Eurasia during the Pleistocene was the saiga antelope, actually a caprine rather than a true antelope. These herbivores form enormous herds and make long seasonal migrations. They have unusual looking humped-up inflatable noses that filter the dust of the steppe in the summer and warm the air before it reaches their lungs in the winter. Their population was reduced to several hundred individuals by 1930, and they were in danger of extinction. Luckily, effective measures were taken for their protection and there are now more than two million saigas ranging from the western shores of the Volga to Mongolia.

Although the climate north of the Pyrenees was very continental and cold during the glaciations, the Iberian Peninsula was, then as now, almost an island. The ocean waters that almost surround it never froze, although they did recede somewhat. But the continental shelf that surrounds the peninsula is very narrow and sharply inclined, so that the 400 foot drop in the sea level did not bring about a major retreat from the shoreline that we know today. Still, the islands of Mallorca and Menorca were joined, and at various points on the Mediterranean the shore was more than twenty miles from where it is now. The bathymetric curve representing a depth of 300 feet is far from the shore, more than thirty miles off the plain of Valencia. That is why some cave sites that are found today on seaside cliffs actually overlooked a broad coastal plain during the glaciations. Nevertheless, the overall area of the peninsula was scarcely greater than it is today, so continentality did not sharply increase in the interior. But it is primarily due to its more southerly position that the peninsula was not as severely dry and cold as Central Europe, the Low Countries, or Northern France.

Current range
of the saiga

Figure 17: Maximum range of the saiga antelope during the last glaciation (lighter area).

The woolly mammoth, the woolly rhinoceros, the reindeer, and the saiga antelope entered Europe from Siberia and Central Asia. They are typical species of the last glaciation, although at some sites we may also find their remains from earlier periods of cold. The case of the musk ox is very interesting. It seems that animals of this type lived on the Eurasian steppes during the

Pleistocene and adapted themselves to the cold during the last glaciation, becoming the arctic species that we now know. The same may have occurred in the case of the arctic fox.

In addition to fossils, we have artistic representations to tell us about the animals in our ancestors'ecosystem. Representations of animals are painted and engraved on rocky cave walls and as portable art on stone plaques and organic materials such as bone, ivory, and antler. All were produced by Cro-Magnons. The really stirring thing about Paleolithic art is that it enables us to see the mythical mammoth, the powerful rhinoceros, the fearsome lion, and the huge cave bear through the eyes of prehistoric humans. The paleontologist Björn Kurtén once wrote that his science, our science, wasn't the study of beings that had *died* long ago, but rather of beings that *lived* long ago. To a paleontologist, it is always moving to see the giant fossil mammals of the Ice Age depicted *full of life* on the walls of a cave.

France and Spain are privileged to be home to most of the Paleolithic cave art found in Europe. In Spain it is largely concentrated on the Cantabrian coast, but more is constantly being brought to light in caves all around the peninsula. In recent years a large and wonderful set of animal engravings has also been discovered in the open, alongside the River Côa in Portugal, as have other smaller ones at Mazouco, in Portugal, and Siega Verde, Salamanca, and Domingo García, Segovia. Portable art, on the other hand, is distributed much more widely across Europe and as far east as Siberia.

Deer, horses, bison, goats, and aurochs, members of an extinct bovine species also called uru, are abundantly represented in Spanish Paleolithic cave and portable art. Reindeer, chamois, wild boar, mammoths, woolly rhinoceri, and carnivores are also portrayed, but less frequently. The geographic distribution of reindeer is interesting because it is an indicator species of extremely cold conditions and a tundra or tundra-taiga environment. It is so closely associated with the last glaciation that the period has come to be known as the Age of the Reindeer. There are reindeer fossils in various caves along the Cantabrian coast and reindeer figures in cave and portable art in the same region. Reindeer fossils have also been found in Puebla de Lillo, León, and probably in A Valiña, a site in Lugo, Galicia. José Javier Alcolea, Rodrigo de Balbín, and other colleagues have reported a reindeer engraved on the wall of a cave, not coincidentally called Reindeer Cave, at an altitude of 2,800 feet in Guadalajara, and another in a cave called La Hoz at 3,450 feet in the same province. These representations demonstrate that reindeer ranged as far south as the central highlands of the peninsula's interior. The same authors report reindeer images among the open-air engravings at Siega Verde.

Much the same can be said of the woolly rhinoceros, whose fossils have been found farther to the south of the Cantabrian coast. My own brother, Pedro María, has studied a cranium of this species retrieved from a dig at Arroyo Culebro in the province of Madrid. The woolly rhinoceros had two horns, the one in front being very long, sometimes nearly four and a half feet. They were large, comparable in size to today's white rhinoceros. Large males weighed over two and a quarter tons and stood over six feet high. As in the case of the mammoth, we have several mummified woolly rhino bodies that we can study in detail. The woolly rhinoceros is not widely represented in the Paleolithic Spanish bestiary, but there is one figure in a cave at Los Casares, close to the cave at La Hoz, in Guadalajara. According to Rodrigo de Balbín and José Javier Alcolea, there is also one in the open-air assemblage at Siega Verde, and Soledad Corchón has identified another on an engraved plaque found in a cave at Las Caldas, Asturias.

Nineteenth century range.

Figure 18: Maximum range of the reindeer in Eurasia during the last glaciation (lighter area).

Fossil evidence tells us that the woolly mammoth ranged over much of Iberia, as far west as Galicia and Portugal, and as far south as the peat bog at Padul, near Granada at an altitude of 3,300 feet. There are also a few artistic representations of them, including one in a cave at Pindal, in Asturias, in a cave at Castillo, in Cantabria, and several on the plaque from Las Caldas, superimposed on each other and on the woolly rhinoceros mentioned above. There are questionable reports of mammoth figures from caves at Los Casares, Guadalajara, La Lluera I in Asturias, Las Chimeneas and La Pasiega, Cantabria, Ojo Guareña, Burgos, and El Reguerillo in the province of Madrid. The red mammoth in the cave at Pindal is very special because its heart, also red, is depicted inside its chest. The cold weather fauna do not seem to have penetrated deeply into Mediterranean

Iberia, but there are some mammoth, reindeer, and musk ox fossils in Upper Paleolithic strata north of the Ebro River in Catalonia.

The archeologist Pilar Utrilla has found six fossilized bones from the Upper Paleolithic in a cave at Abauntz, Navarra, which Jesús Altuna and Koro Mariezkurrena have recently classified as belonging to saiga antelope. The same Upper Paleolithic stratum has also yielded reindeer remains. Jesús Altuna had previously identified some engravings in a cave at Altxerri, Guipúzcoa as saiga, although other authors had interpreted them as representing chamois. It was known that the range of the saiga antelope had approached the Iberian Peninsula, because their remains had been found in a dig at Isturitz, in the French Basque province of Basse-Navarre, and in Dufaure, a little farther north in the southern part of Landes Department. Reindeer and saiga antelope were really quite abundant across the Pyrenees on the broad plains of Aquitane during the last glaciation. In fact, since the six fossilized bones are five phalanges and a centrotarsal, another foot bone, Jesús Altuna and Koro Mariezkurrena speculate that a human may have carried them to Abauntz as part of a pelt. After all, the terrain around the cave itself is rough and broken, not a likely habitat for these animals, who are lovers of the open spaces found on plains and expansive steppes.

When mammoth, woolly rhinoceros, and saiga antelope remains are found in archeological contexts on the Iberian Peninsula, they are always at Upper Paleolithic strata, corresponding to the presence of modern humans. But Jesús Altuna, who has been conscientiously studying the fauna at archeological sites in the Basque region and on the Cantabrian coast in general for years, has identified reindeer fossils at the strata of Mousterian Neanderthal remains, one each at Axlor, in Vizcaya; Lezetxiki, in Guipúzcoa; and Abauntz, in Navarra. In addition,

17.3 inches

Red mammoth at the Pindal cave

Figure 19: Maximum range of the woolly mammoth in Eurasia during the last glaciation (lighter area).

Obermaier reports four reindeer remains at what seems to be the level of the penultimate glaciation at the cave in Castillo.

The Cantabrian fossil record includes some remains that have been identified as belonging to the spectacular cervid *Megaloceros giganteus*, found at strata corresponding to both Neanderthal and Cro-Magnon populations. The megaceros was distinguished by its great size and the enormous size of its palmate antlers, with a span that could exceed twelve feet and a weight of nearly a hundred pounds. Large males would not have been able to enter a forest carrying such huge structures atop their heads, for their antlers would have been caught in the tree limbs. It is reasonable to think that they must have lived in more open environments, probably cool ones. So many megaceros fossils have been retrieved from Irish bogs that they have come to be known by some people as "Irish elk," although they have no especially close relationship to today's elk. Others call them the "giant swamp deer," although they were not deer like those of today. This species of megaceros and other earlier ones seem to have been present throughout the Iberian Pleistocene, though never in great abundance. Rodrigo de Balbín and José Javier Alcolea believe that Iberian Cro-Magnons represented the megaceros in a figure at Siega Verde, although further study is needed to confirm their interpretation. In any case, the species became extinct toward the end of the Pleistocene.

In addition to the herbivores we have mentioned, there were roe deer in Iberia during the last glaciation. Their presence is a sign of a forest habitat. There was also an extinct equine species, *Equus hydruntinus*. It was smaller than a horse, closer in size to a donkey, but probably unrelated to it. And lest we forget the other European primate, remains of the Barbary macaque have been found pertaining to this epoch in a dig at Cova Negra, in Játiva. It too was present before the most frigid period of the last glacial maximum. The macaque is a Mediterranean species and was present in many other parts of Europe throughout the Pleistocene as well, ranging as far north as England and Germany, but always during interglacials. The macaque and many other "interglacial species," the hippopotamus, for example, should naturally be recolonizing Europe in this warm Holocene that we are presently experiencing. But I don't expect that we will let them live among us.

Some of the carnivores that lived during the last glaciation are familiar to us, like the wildcat, the lynx, the red fox, and the wolf. Few people in Europe know anything about the dhole, a relative of the latter two. But for some of us, any mention of the dhole brings back fond memories of our lost childhoods. Dholes appeared in a dramatic chapter of *The Jungle Book* by Rudyard Kipling. In this chapter, Mowgli's wolf pack engages in a fierce battle with a pack of dholes, or red dogs, as they are called for their distinctive color. I still remember my grief at Akela's death in combat. He had been Mowgli's protector and the beloved leader of the Seeonee.

Although dholes are found in Mousterian Neanderthal and earlier digs, they seem to have become very rare or even extinct on the peninsula before the end of the Pleistocene. The only find I know of during the time of the Cro-Magnons is at Amalda, in Guipúzcoa, at an Upper Paleolithic level that also contains reindeer. Today the dhole lives only in Asia. Though they are smaller than any wolves, a pack of dholes is a ferocious thing.

The presence of the polar fox is a definite indication of frigid conditions, but it is hard to distinguish this species from the common red fox on the basis of a few bones or teeth. Even so, Jesús Altuna has identified the remains of a polar fox at the same level of the Amalda site that contains dhole and reindeer.

There is plenty of evidence of the presence of gluttons on the peninsula. I am not referring to people with inordinate appetites, but to the wolverine, another name for the same animal. The wolverine is the largest species among the mustelids, a family that also includes the pine marten, the stone marten, the weasel, the ermine, the mink, the badger, and the otter. The wolverine lives only in the far north today, from Scandinavia to Canada, but its remains have been found at an Upper Paleolithic level of a dig at Lezetxiki that has also yielded woolly rhinoceros, and at a dig without an identifying archeological context at Mairuelegorreta, in Álava. It is also possible that an engraving at the Los Casares cave in Guadalajara represents one of these Nordic creatures. An ivory sculpture of an animal head that Jesús Jordá believes to be a wolverine has been found in Jarama II, a cave in the high valley of the Jarama, also in Guadalajara. If he is right this would be the representation of an extinct cold weather species, the wolverine, sculpted into the fossil remnant of another cold weather species, the mammoth.

The cave at Los Casares is a treasure trove of evidence about life on the highlands at the center of the plateau during the Upper Pleistocene. We find a Neanderthal dwelling place along with varied fauna representing a cool climate in a medium altitude environment. The fauna include marmots, beaver, boar, red deer and roe deer, chamois, horses, goats, steppe rhinoceri, wildcats, lynx, leopards, lions, fox, wolves, dhole, both brown and cave bear, spotted hyena, and a bovine species, either aurochs or bison. A human metacarpal has also been found, from the right little finger of a Neanderthal. The Neanderthals later disappeared from the scene and after a lapse of 15,000 years or more, the Cro-Magnons arrived. They painted and engraved some of the animals of the day on the walls of the cave: horses, aurochs, deer, goats, a woolly rhinoceros, the possible mammoth, the possible wolverine, and a large cat, possibly a lion or lioness. There is also a series of anthropomorphic figures, very distorted representations of the human form. The woolly rhinoceros, mammoth, and wolverine, if the identity of the latter two can be confirmed, are clear evidence of a cold environment on the plateau at the time of the Cro-Magnons, and of steppes upon which large herds of herbivores would have grazed. But the presence of deer also indicates the existence of a forest habitat, possibly on the valley floors. Perhaps the rugged topography of Spain can again explain the association of seemingly incompatible species. Although no reindeer have been found at Los Casares, one is represented at the very nearby and even higher cave at La Hoz, altitude 3,450 feet.

Both the Neanderthals and the Cro-Magnons coexisted and competed with leopards and lions. The feline representations are not very good in Paleolithic Spanish art, but anyone who wants to see strikingly realistic images of lions dating from that period need only go see the paintings at the Chauvet cave in France. In every known case, including the images at Los Casares for example, the lions are represented without manes. This may be because our prehistoric artists wanted to depict only females, or because the males of the lion species living in glacial Europe did not have the manes that adorn African and Indian lions today. In the coldest regions, lions became very large for the same reason that the woolly mammoths had smaller ears than today's elephants: the relationship between body surface and volume. In a cold environment, a warm-blooded animal loses body heat through the skin, so it is in that animal's interest to have the smallest possible surface area relative to size. Paradoxically, this can lead to overall growth. A few simple figures can demonstrate how increased body size will affect the regulation of body temperature. A cube with one foot sides has a volume of one cubic foot and a surface area of six square feet. A cube with two foot sides, on the other hand, has a volume of eight cubic feet, eight times greater, while the surface area is twenty-four square feet, only four times greater. In other words, the relative surface area is only one half for the cube with two-foot sides what it is for the cube with one-foot sides, and the theoretical heat loss would likewise be reduced.

Neanderthals and Cro-Magnons are popularly thought of as cave dwellers, "cavemen." But the cave bear is so deserving of its name that it is incorporated into the scientific nomenclature. It is *Ursus spelaeus* in Latin, and "cave bear" is its literal translation. These animals reached huge proportions, greater than the brown bears we know today. The average brown bear, considering both sexes and all of their various populations, weighs something over 350 pounds, although our few remaining Cantabrian bears and the nearly extinct Pyrenean bears are much more modestly proportioned, the very largest males rarely surpassing 425 pounds. The largest brown bears in the world today are the grizzlies of Alaska and British Columbia, especially those on Kodiak Island in the Gulf of Alaska. After gorging on salmon, they may weigh nearly 900 pounds. But the *average* male cave bear weighed about a thousand pounds and the female somewhat less, but certainly over 650 pounds. Their height at the withers was only about four feet, not very tall, since these long extinct animals were plantigrade and prodigiously corpulent.

Lions. Chauvet Cave
(32,000 years)

Figure 20: On the left, lions painted on the walls of Chauvet Cave.

Cave bears hibernated in caves as modern brown bears do. Over the ages, large numbers of them died during hibernation, so more than a few underground caverns hold hundreds or even thousands of their skeletons. Despite their imposing appearance, cave bears were not impressive hunters. Their huge teeth were better suited to chewing fruit than to cutting through flesh. Their large canines had become blunt, employed as they usually were in tasks other than killing prey. Still, it would not have been much fun to compete for "housing" with these giant plantigrades.

The cave bear was an almost exclusively European species, living both in the continent's temperate forests and on its colder steppes. But it does not seem to have inhabited the Mediterranean environment. On the peninsula, they have been found only in the Pyrenees, on the Cantabrian coast, in Galicia, and on the two major plateaus. The southernmost site where they have turned up is the cave at Reguerillo, in the province of Madrid. Although their fossils are not as abundant, brown bears coexisted with the cave bears and their ancestors in the European Quaternary. A pair of them, one headless, is painted on a wall of the cave at Ekain, in Guipúzcoa. Another very attractive individual appears in the cave at Santimamiñe, in Vizcaya. Brown bears are easily distinguishable from cave bears. The anterior limbs on the latter species are much longer than the posterior ones, so that the line of their backs descends more abruptly from the withers to the rump. There is a magnificent cave bear engraving in the cave at Venta de la Perra, in Vizcaya. With the end of the last glaciation and the ice age, the cave bear no longer presented any competition to the brown bear, disappearing forever into the mists of time.

Finally, both Neanderthals and Cro-Magnons competed for carrion with the fearsome spotted hyena. Like humans, the hyena can be very effective when hunting in a cooperative group. The spotted variety grew to be quite large during the last glaciation. While the striped hyena was less of a threat to the humans, it was still a strong competitor for carrion. Perhaps we can say that they were equally dangerous to the humans' stomachs if less so to their physical safety. They seem to have been less common on the peninsula than the spotted hyenas. They have been found only at the Mousterian dig at Furninha, in Portugal.

Atapuerca: The Magic Mountain

So far, we have reviewed the large mammal species of the last glaciation, those that coexisted with the Neanderthals and the Cro-Magnons. But what about the herbivores and carnivores at the time when the Neanderthals' ancestors first populated Europe? There is no better way to become acquainted with them than to return to the Sierra de Atapuerca in Burgos, where a team of Spanish paleontologists and archeologists is bringing to light an extraordinarily important fossil record that will cover most of that period.

The Sierra de Atapuerca is a great mass of limestone. This limestone was formed at the bottom of the sea more than 85 million years ago during the Cretaceous, the last period of the Second, or Mesozoic Era, the time of the dinosaurs. Later, in the Oligocene Epoch of the Tertiary Period of the Cenozoic Era, the "Age of Mammals," the immensely powerful forces that move the earth's crust thrust the limestone from the sea, violently deforming it. It became a small mountain, and like other mountains produced through the same process, it is known as a recumbent fold or anticline. Once the peak of the mountain was definitively out of the reach of the sea, the forces of erosion began their work, eventually reducing it to its present flat aspect and its maximum altitude of 3,550 feet above sea level. Then, in the following Miocene Epoch, what we know today as the *meseta* of the River Duero, or the Duero Plateau, became a giant continental basin with no outlet to the sea. It slowly filled with sediment eroded from the mountains that surrounded it like a great wall. These were and are the Cantabrian Range to the north, the

Iberian System to the east, the Central System to the south, and the León Mountains and the Tras Os Montes to the west. The esteemed Don Eduardo Hernández-Pacheco used to compare the Duero basin, i.e. the high plateau of Castile, to a broad parade ground within an enormous castle whose defenses were a ring of mountainous bastions.

Atapuerca is located at the northeast corner of the great Duero basin, just a few miles from the Sierra de la Demanda, a part of the Iberian System. It sits immediately alongside the La Bureba Corridor, one of the mountain passes that open to the interior of Castile. The other passes are in the southeast corner, in the province of Soria, and in the southwest, the pass at Ciudad Rodrigo, whence the Duero flows into the Portuguese territory of Los Arribes and ultimately to the sea. A little beyond the Sierra de Atapuerca, past the Pedraja Pass, at about 3,700 feet, one reaches the valley of the Ebro. The venerable Camino of Santiago follows this natural path between the two river valleys, which says a lot for the strategic position of the Sierra de Atapuerca, and probably helps explain the intensity and continuity of the human presence in these Castilian lands.

So much sediment accumulated in the Duero basin that by the end of the Miocene, the Sierra de Atapuerca barely peeked out above the surrounding plain. Additional limestone, this time continental rather than marine in origin, was then deposited in the shallow lakes that were dispersed throughout the basin. This limestone that was deposited at the final stages of the sedimentary period did not undergo any folding process. Today it forms a horizontal platform, an upland formation known as a *páramo*, that reaches almost to the top of the Sierra de Atapuerca.

The filling-in of the Duero basin stopped in the Pliocene, the epoch that followed the Miocene, when uplift at the center of the peninsula created a river network that began to erode and carry away the sediments that had accumulated for millions of years, depositing them in the Atlantic, a process that continues to this day. Throughout the river basin, the river's waters have cut through the limestone surface of the *páramo* and exposed the softer clay and marl sediments below. Thus the Castilian plateau has two levels with very different characteristics: the older and higher surface of the *páramo*, generally consisting of limestone and very little topsoil, and the newer, more fertile surface of the inhabited valley floors. Steep cliffs called *cuestas* separate the two levels of the plateau.

These topographical features, the *páramo* and the *cuesta,* can be seen on the left bank of the River Arlanzón at the town called Ibeas de Juarros, directly across the river from the Sierra de Atapuerca. The Arlanzón flows close by the southern slopes of Atapuerca just a few miles upstream from the city of Burgos. Like all rivers, it pries loose and drags large rocks from its upper reaches, which it then smoothes and shapes into rounded boulders. When the river is highest, many of these boulders are deposited on its flood plain. As I have said, the river has slowly exposed more and more of the soft clay and marlstone sediments that filled in the Duero basin during the Miocene. But remnants of the older and higher stone surface are also present. In geological terms, they are called *terraces*. At their highest they are 280 feet above the current level of the river, at about 3,260 feet above sea level, not far below the high point of the Sierra. By examining these terraces we have been able to determine the prehistoric course of the River Arlanzón. We discovered that the caves of the Sierra de Atapuerca were very close to the river bank at the time when prehistoric humans were active in what today constitute our archeological sites. We can imagine our prehistoric ancestors on the hillsides, gazing down upon herds of herbivores grazing peacefully in the valley of the lazy Arlanzón below, and upon the Valhondo trough, through which its tributary the Pico, though much diminished, still flows today.

The marine limestone that constitutes the substrate of the Sierra de Atapuerca was readily dissolved by water. When this occurred, long subterranean channels were formed, and water circulated through them under pressure, creating and continually extending a network of channels known as a karst. The phreatic surface, or water table, dropped because the river system kept eating its way deeper into the limestone. This left the uppermost chambers of the karst dry, their roofs collapsed, and they were left open to the surface. As the hillsides were eroded by the river, lateral openings to the caverns were produced. At that point, carnivores and humans gained access to them for use as cave shelters.

The three most active digs so far are called the Gran Dolina, the Galería, and Sima de los Huesos. They are close to each other, particularly the first two. They contain fossils between a quarter million and a million years old. In addition, both older and younger fossils are currently being explored in other caves on the Sierra de Atapuerca, and we are optimistic about what they may have to tell us. The Sima del Elefante contains the two oldest strata, and the Mirador cave, as well as the first chamber of Cueva Mayor, known as the Portalón, contain the most recent ones. The paleontological and archeological treasures of the Sierra, far from being exhausted, are revealing more of themselves each day. The paleontologists studying animal fossils at these sites are the rodent specialist Gloria Cuenca, the carnivore specialist Nuria García, and our herbivore specialist Jan van der Made. Together with the paleobotanist Mercedes García Antón, they will now guide us through the ecosystems of the Sierra de Atapuerca in its remote past.

In order to divide the fossil record of our digs into two major periods, I will refer you to the evolution of a rat, or more specifically, a water rat. Little more than 500,000, probably about 600,000 years ago, a species of water rat called *Mimomys savini* became extinct and was replaced by the species *Arvicola cantianus*, very similar to today's water rats, which by the way are barely related to our urban dwelling gray rats, beyond both being rodents. All other fossils that have been found together with the species *Mimomys savini* are *ipso facto* more than half a million years old. *Mimomys savini* is found from the third level to the lower part of the eighth level at the Gran Dolina, a vertical distance of some twenty-eight feet through sediment representing the period from a little more than half a million to almost a million years ago. No fossils are found below level three. I should explain that the numbering of levels from the bottom up at the Gran Dolina is exactly opposite to the norm. This site was initially opened inadvertantly, exposed by the excavation of a railway trench for a mine at the end of the nineteenth century. As a result, the stratigraphy of the site was exposed immediately and in its entirety, eliminating the usual process by which scientists work downwards, carefully identifying and enumerating one layer at a time.

A great variety of large mammals lived in the ancient Sierra de Atapuerca, a spectacular assortment compared with what we find today. Beginning with the herbivores, there were large two-horned rhinoceri of the species *Stephanorhinus etruscus*. There were boars, horses, red deer, roe deer, and probably fallow deer. There were primitive megaceros of the species *Eucladoceros giulii*. A magnificent bovine cranium of the species *Bison voigtstedtensis* has been recovered from the Gran Dolina, and from level seven we have the back feet of an ancient musk ox, an ancestor or at least a relative of today's musk oxen. As has been noted, the musk oxen of that time were residents of the steppes, not yet having adapted to periglacial conditions. To complete our picture of the herbivores that human Atapuercans could have spied from their cave mouths, we should include a group of hippopotami swimming in the River Arlanzón or one of its tributaries, where beavers also constructed their dams. Strange as it may seem, hippopotami inhabited the peninsula right up to the cold of the last glaciation, and beavers have never left Europe, although they did die out in Iberia.

In addition to the beaver, two other large rodent species are found at the oldest levels of the Gran Dolina, the porcupine *Hystrix refossa* and the marmot. All but one of the closest relatives of *Hystrix refossa* live in warm climates today in Africa and Asia, while one species is found in the Balkans, in Sicily, and part of the Italian peninsula. It may have been introduced by ancient humans. The porcupine was a common species in the European Pleistocene, particularly in warmer areas and during warmer periods. Marmots live today in the Alps and in the Tatry Mountains on the Polish-Slovakian border. They have also been very successfully reintroduced to the Pyrenees. They live on high mountain meadows and hibernate in burrows. It is possible that at some very cold point between 600,000 and 900,000 years ago, the highest part of the Sierra de Atapuerca may have been completely treeless, a congenial habitat for marmots. On the other hand, a bird of prey such as an eagle or an eagle owl may have captured a marmot high in the nearby Sierra de la Demanda and brought it back to a nest in the Sierra de Atapuerca, and it is even possible that a terrestrial carnivore may have brought one back.

Now let us consider the predators in these ancient ecosystems on the Sierra de Atapuerca. Leaving aside for a moment the question of what place humans occupied in the food chain through which vital material and energy circulated, the preeminent predator was a large felid with swordlike teeth, the *Homotherium latidens*. This saber-toothed tiger was comparable in size to a lion and had enormous curved upper canines, or fangs, finely serrated on two edges. Although the species disappeared from Europe half a million years ago, its cousin species *Homotherium serum* survived in the Americas until the end of the ice age. We have no definite answer as to how *Homotherium latidens* and other related sabertooths, referred to collectively as Machairodontinae, used their elongated upper canines. Some authors believe that they used them like daggers, plunging them into their prey in order to bleed them to death. Others believe that they used them to puncture the skin and adjacent tissues of their victim's abdomens, and then, tossing their heads back, they would wrench away a mouthful of the prey's body. Even if their prey escaped after this treatment, the cats would have only to follow them and await their inevitable death from loss of blood. This is hypothesized as a technique that would have enabled the cats to prey upon animals much larger than themselves, young mammoths for example, by attacking, inflicting a mortal wound, and then retreating to bide their time while their victim died.

Alan Turner and Mauricio Antón, on the other hand, believe that the stresses imposed on the animals' canines by the above-mentioned uses would have seriously damaged them. Given their canines' importance to the procurement of food, the cats would then surely have perished. These two authors find it more credible to suppose that after taking down and immobilizing its prey, a big cat would have used its canines to perforate its victim's throat and cause death by asphyxia, or to open one of the neck's large blood vessels, as today's lions do when they bring down a large herbivore. These uses would not have presented so much potential for irreversible damage to the teeth.

Another large felid of the earliest period of the Sierra de Atapuerca was the European jaguar, or *Panthera gombaszoegensis*, extinct since about 400,000 years ago. It was the size of a modern American jaguar: smaller than *Homotherium*, but larger than a leopard. The lynx is another, even smaller, feline whose remains are found at the lower levels of the Gran Dolina. It seems that cats of all sizes, probably ocelots as well, inhabited the ecosystems of the Sierra de Atapuerca, *Homotherium* being the largest of all. Lions probably first appeared in Europe some 600,000 years ago and the sabertoothes vanished from the continent shortly thereafter, most likely having lost out to the lions in their competition to be the dominant predator of the ecosystems of that time.

The remains of two canine species have been found, *Vulpes praeglacialis*, an ancestor of the arctic fox that had not yet adapted to the periglacial environment, and *Canis mosbachensis*, a small wolf not much bigger than a modern jackal, which over the course of 400,000 years would grow larger and evolve into the wolf as we know it today.

The lower levels of the Gran Dolina have yielded Europe's oldest remains of the spotted hyena, a social carnivore and a strong competitor of humans for both carrion and live prey. The spotted hyena had specialized teeth for breaking open the bones of large herbivores to extract marrow, something humans could also do, but only with the aid of a hard object to strike and crack them open. Up to this time, however, the Atapuerca fossil record does not include any trace of *Pachycrocuta brevirostris*, the largest hyena of all time. Its absence is intriguing. It has turned up in many contemporaneous European sites and in Iberian sites older than the Gran Dolina, such as at Cueva Victoria, Murcia; at Venta Micena, Granada; at Incarcal, Gerona; and at Pontón de la

Oliva in the province of Madrid. Nuria García believes that when the spotted hyena arrived, it began to replace *Pachycrocuta brevirostris*, first in southern Europe, as Atapuerca seems to indicate, and later in the rest of the continent, until the latter species disappeared some 400,000 years ago.

Numerous bear remains have also been found at these lower strata of the Gran Dolina, identified as belonging to an old species that may have been a primitive form of the brown bear or perhaps of the cave bear.

We also have many fossils of later species at the Galería, at Sima de los Huesos, and in the uppermost parts of the Gran Dolina, from level eight to level eleven. Taken together, these remains represent a period from a quarter million to almost half a million years ago. Throughout the period we continue to find horses, fallow deer, red deer, megaceros, bison, and rhinoceri. The scarcity of fossil material makes it impossible to assign the bison remains to a specific species. They could have been *Bison schoetensacki*, often called the forest bison, or *Bison priscus,* the larger steppe bison. Despite these adjectives, the bison fossils are not useful ecological indicators, because both species would typically roam from one environment to the other. Some of the recovered bovine remains may in fact be from the aurochs, *Bos primigenius,* or even from water buffalo, which are exclusively Asian today, but which once inhabited Europe. It is difficult to distinguish between the various bovine species based solely on bones found in isolation from their skeletons of origin.

The rhinoceros *Stephanorhinus hemitoechus* inhabited Atapuerca at this time. They grazed on the steppes and coexisted in Europe for a long time with the larger Merck's rhinoceros, *Stephanorhinus kirchbergensis.* The latter was a truly impressive animal, taller than any of today's rhinos at a height of up to eight feet. The two species, Merck's rhino and the steppe rhino, were able to coexist because they were adapted to the exploitation of different resources. Each had its own ecological niche, which reduced the competition between them. Merck's rhinoceros browsed on the tender parts of woody plants, like leaves, buds, and fruit, and was therefore more of a forest dweller. Two rhinoceros species also coexist within a single region today in Africa, where the black rhino browses on the tender parts of trees and brush while the white rhino grazes on open grasslands. In any case, we are not sure that Merck's rhinoceros was ever really plentiful on the peninsula. Both it and the steppe rhinoceros were adapted to temperate conditions and disappeared from central Europe at the onset of the last glaciation. That was the time of the Neanderthals. The steppe rhino held on a little longer on the peninsula, until the cold intensified and the woolly rhinoceros was fully established and dominant. This was now the time of the Cro-Magnons.

Among the large rodents, we continue to find large marmots and porcupines in the second epoch explored at Atapuerca, although the porcupines are of a different species, *Hystrix vinogradovi*. A number of carnivores are found at Sima de los Huesos. *Ursus deningeri*, an ancestor of the cave bear, is represented best of all. There are also wolves, foxes, wildcats, and a lynx species on the evolutionary line of today's Iberian lynx. Lions have also turned up, and there is an enigmatic feline fossil, a metatarsal, one of the bones of the foot, that based on its size may have originated with a leopard or a European jaguar. That means that there are felines of four different sizes. The small carnivores known as mustelids are often overlooked, but they are represented at this site by two species, a large one similar to a marten or a pine marten, and another, smaller one, similar to a weasel or an ermine. The Galería has also yielded remains of dhole and badger.

It is surprising that hyenas are absent from the Sima de los Huesos, the Galería, and the higher levels of the Gran Dolina so far excavated. Nuria García has proposed that humans may have driven them out of the Sierra de Atapuerca as unwanted competitors. Perhaps we will find hyenas at older levels of the Gran Dolina. Earlier humans may have been unable to overpower them, being fewer in number or less well organized, or they may have spent less time in the immediate area. But hyenas abounded on the peninsula in periods following those represented by the Atapuerca finds. We can hypothesize that as time went on, humans became more active hunters and less interested in carrion. Then they would have diverged from the ecological niche that they had long disputed with the hyenas and largely surrendered it to the latter, who seek carrion more than they hunt.

Only two sets of elephant remains, indeterminate as to species, have been recovered at Atapuerca so far. One comes from the Gran Dolina and the other from the Sima del Elefante, from whence its name. This is not to say that there were no elephants in the vicinity during the long period of time represented by the paleontological explorations thus far. The straight-tusked elephant *Palaeoloxodon antiquus* was abundant in Europe, like the hippopotamus, a typical faunal species of interglacial periods. Many remains of this elephant, up to twelve feet tall, have turned up on the peninsula. Perhaps the best known are those of Torralba and Ambrona, in Soria, of which more later. But the straight-tusked elephant disappeared from Europe during the glaciations, or at least from its colder regions, giving way to the steppe mammoth, *Mammuthus trogontherii*, forerunner to the woolly mammoth. *Mammuthus trogontherii* was the largest European elephant of all time, reaching nearly fifteen feet and weighing more than eleven tons. Like the steppe rhinoceros and Merck's rhinoceros, the last straight-tusked elephants seem to have survived on the peninsulas of the Mediterranean until the definitive cold that brought the woolly mammoths and rhinoceri into these ecosystems and provided the kiss of death to the European hippopotami.

Let us take a short look at plant life. Oaks, gall oaks, holm oaks, and cork oaks all belong to the genus *Quercus*. The leaves of the first two wither in autumn, while those of the second two are always green. Both of these groups, and others as well, show up in the Sierra de Atapuerca pollen record. The pollens tell us that the forests were not very different from what is found today both there and in the nearby Sierra de la Demanda. But sometimes we find pollens that indicate the presence of species endemic to a more Mediterranean climate, such as the carob tree, the European hackberry, the wild olive, the narrowleaf phillyrea, and the mastic tree. At other times the forests were dominated by cold weather pines and gymnosperms, the family of trees that includes the cypresses and junipers.

The fossil record clearly tells us that animal and plant life in the Sierra de Atapuerca was extraordinarily diverse throughout the Pleistocene. But no single ecosystem can explain the existence of such a wide variety of species. This was probably due to the great variety of habitats found in the Sierra and its environs, including broad grasslands, waterways, limestone outcroppings, and the lofty Iberian peaks nearby. Remnants of natural forest still survive today here and there. The land atop the Sierra de Atapuerca itself has never been plowed, and since timbering ended there, limestone-loving holm oaks and gall oaks are spontaneously regenerating a forest on its thin topsoil. The slopes of the Sierra have fared less well. The Pyrenean oaks that once grew on its terraces and Tertiary soils have almost all given way to the plow and cereal grain agriculture. That is why it is so vital to defend this evocative island of natural history at all costs. It is an impressive example of Spanish patrimony less than nine miles from the city of Burgos. It must be protected for scientific and pedagogical reasons as well. The excavations there, and the items retrieved from them, are an invaluable treasure, not only for us but for generations to come. But our existing discoveries and others yet to come will be meaningful only to the extent that they can be understood in the context of their natural environment.

The Sierra de Atapuerca is unique because it closely documents climatic and ecosystemic changes, aspects of human evolution, and the development of human behavior and technologies over at least the last million years, an extremely long period of time, longer in fact than any other place in the world. The oldest human fossils in Europe have been found here. With them, we have been able to identify an additional human species, the oldest known case of cannibalism, and the first known funerary practices. This extraordinary paleontological and archeological record makes it one of the outstanding site complexes in the world. It is a truly magic mountain, the most important historical site in Europe and in a profound sense, the heart of Spain.

CHAPTER SIX

The Great Extinction

For the next few days we traveled along the southern edge of the Mann Ranges. To most white men that country is desolate, unfriendly, "the land that God forgot." The aboriginal's outlook is the reverse. To him, the landscape abounds with interest. The trees are not just trees, but the transformed bodies of bygone heroes; creeks are not just places where the water runs, but the tracks of gigantic serpents that meandered their way across the country...

Charles P. Mountford, Brown Men and Red Sand:
Journeyings in Wild Australia

One Strong Sex or Two?

When we discuss the social economy of prehistoric humans, we say that they were hunter-gatherers, that they lived off the food available in their environment. I'm not referring here to the first African hominids (the *Ardipithecus*), the Australopithecines, or the paranthropoids, who are believed to have been almost exclusively vegetarian. Nor do I refer to *Homo habilis,* the first hominid to add meat to its vegetarian diet. I am thinking rather of those of our ancestors who were physically like us, the true humans, from *Homo ergaster* onward, who emerged in Africa perhaps two million years ago and subsequently populated Eurasia.

It is not difficult to understand the role of the hunter. Europe was brimming with animal life during the Pleistocene. There was an abundance of large herbivores, including goats, deer, horses, oxen, bison, rhinoceri, and elephants, all potential prey for any predator. But it is not clear whether prehistoric humans hunted these animals, or if their diet included only the meat of animals that had died naturally or that had already been brought down by other predators. They may have had less in common with the lion than with the hyena, a carnivore held in some disrepute, never to be selected as the symbol for any marketing campaign or the logo of any sports franchise or military unit. Truth be told, the hyena is an impressive hunter in addition to being a carrion eater. But one gets the impression that humans would generally prefer not to be identified with this four-legged creature famous for its maniacal laugh.

We will discuss the hunter/carrion-eater conundrum later on. But let us now turn our attention to another, no less important, puzzle, the disjuncture represented by the dual archetypes of male hunter and female gatherer. Since hunting is envisioned as an eminently male activity, the traditional view that hunting provided the bulk of the prehistoric human diet assigned a strongly protagonistic role to "the stronger sex." It is certainly true that the male was and is physically larger and stronger than the female, but that is not to say that his activities contributed more to the group in the way of calories. In fact, the gathering of vegetable products may have represented the real foundation of the prehistoric human diet. We may have to revise our image of the proud hunter returning to the shelter loaded down with his kill and finding his female partner, numerous offspring, and perhaps a parent (or an in-law!), all of them dependent on his success, greeting him ecstatically at the threshold. Berry picking does not require great strength, and everyone can pitch in, including women, the elderly, and even recently weaned children.

How different it would be to picture that hunter returning home empty-handed and having to rely upon the food collected by the weakest members of the family group! This radical change in perspective, a subversion of the traditionally imagined schema of prehistoric life, should make us take a good look at ourselves. Have women been the stronger or more productive sex in terms of obtaining food? Perhaps natural selection has produced some special characteristic exclusive to women and not present in the females of other primate species.

If we compare the human life cycle with the life cycles of our closest living relatives, the two species of chimpanzees, we notice some important differences. Our developmental stage is much longer, our growth is slower, and our first offspring arrive later. The female chimpanzee first gives birth at thirteen or fourteen years of age, while young women among the Ache of Paraguay and the !Kung of Botswana, two modern hunter-gatherer peoples, bear their first children between the ages of seventeen and nineteen. The death knell tolls later for humans as well. Chimpanzees do not live much beyond forty years, while people over sixty are not uncommon among the Ache and the !Kung.

Given this data, it would seem that each stage in our life cycle is uniformly longer in comparison with the chimpanzees'. But this is not the case. Female chimpanzees are fertile practically until the day they die. Or to put it another way, their reproductive apparatus undergoes exactly the same kind of deterioration as do their other bodily systems towards the end of life, a set of circumstances that we call old age. Women, on the other hand, become naturally sterile long before becoming physiologically "old." Among the modern hunter-gatherers that we are looking at, forty percent of all women may be menopausal or post-menopausal. Actually, the duration of fertility among female chimpanzees, gorillas, and humans is not very different, averaging less than thirty years; but among the primate females, only human women have a long postreproductive existence.

Kristen Hawkes, James O'Connell, and Nicholas Blurton Jones think that menopause developed so that mothers could help their daughters to provide food to their grandchildren. This has been called "the grandmother hypothesis." This hypothesis envisions an adaptation based on the genetic economy. That is, it would be more advantageous for a woman past her period of fertility to promote her lineage through her grandchildren, whose genetic makeup is twenty-five percent hers, than to continue bearing children bearing fifty percent of her genes, since she may not have the strength or resources to raise the latter. In addition, the offspring of an older woman could easily be orphaned before they were able to fend for themselves, given the extended duration of human childhood and adolescence. Although we have a potential life expectancy of almost 100 years, it is difficult to imagine that many prehistoric women could have conceived a child at the age of fifty and lived long enough to raise it to maturity.

Only human mothers share food first with their children and then with the children of their daughters. Among other species, chimpanzees for example, females share food only with their direct offspring and take no interest in their grandchildren. As I mentioned in the last paragraph, the grandmother hypothesis proposes that grandmothers would have provided food to the children of their daughters. But it does not suggest that they would have done the same for the children of their sons. This is because the grandmother can be sure that her daughters' children carry her genes, but may not be so certain about the children of her "daughters-in-law." Those children could have been fathered by men other than her sons. As for her "daughters-in-law," let *their* mothers help them!

In order to accept this hypothesis, one must ask two questions: 1) can a grandmother's help really be that crucial to the survival of her daughters' offspring? 2) why don't we see this behavior among grandmothers in other primate species, for example chimpanzees? The answer to the first question is that a grandmother's help could be vital during weaning, a critical period of childhood when the young lose the benefit of their mothers' milk, but are still unable to fend for themselves. Maternal milk not only nourishes them, but also provides protection against infection. A grandmother's help at this critical juncture *could* increase the chances for survival of recently weaned children. In addition, it could allow for earlier weaning, which would shorten the interval between births, effectively increasing her total number of descendants.

This explanation seems reasonable and provides an explanation for menopause, or rather, for why the duration of fertility has been stable in human females even as their longevity has increased. In fact, the authors mentioned above see it the other way around. They conclude that longevity increased in order to provide women with a post-menopausal stage of life. This would mean that men have benefited indirectly from the process, since the genes selected to prolong the female life span would also have been transmitted to the males of the species, increasing their longevity as well.

How can we explain the evolutionary development of menopause? If longevity increased first and then the reproductive period grew shorter, two evolutionary steps would have been necessary. In reality, it seems that fertility among female chimpanzees also ends several years before their natural deaths, if not several decades as is the case for our species. If the common ancestor of chimpanzees and humans carried a gene limiting female fertility to a thirty year period, only one evolutionary step would have been necessary in order to produce menopause in modern women, increased longevity with no change in the duration of fertility. In evolutionary biology, the most economical hypothesis, i.e. the one entailing the least number of evolutionary steps, is considered the most plausible. This criterion is technically known as the "principle of parsimony."

I must confess that I do not go as far as Kristen Hawkes and her colleagues, who believe that if I live to a ripe old age it will be thanks to the evolutionary development of menopause among women. Without reservation, they go even further and assert that the grandmother's contribution to the family economy is so important among hunter-gatherers precisely because it is women, not men, who provide the group with the bulk of their food. Men, they say, have long played a secondary role as providers. In their fieldwork among the Hadza people of Tanzania, these anthropologists have observed that hunting, an exclusively male activity, is a very important source of calories, but too irregular a source to ensure group survival. There are long periods when men repeatedly return from the hunt empty-handed, having neither made a kill nor encountered carrion. If this is the case among the Hadza, these researchers ask themselves, who use bows and poison-tipped arrows and live in an area brimming with large herbivores, wouldn't the situation have been even worse for ancient hunters lacking that modern technology?

The vegetable products harvested for food by the Hadza include many kinds of fruits and underground storage organs, but one in particular is especially important. This is the tuber of the plant *Vigna frutescens*, called *ekwa* in the Hadza language, which grows to a considerable depth and which the Hadza harvest using a simple digging stick. The ekwa is available year-round, including times when both game and other vegetable products are scarce. It is a food that recently weaned children would not have the wherewithal to obtain without adult help, which is in fact provided by their grandmothers. Only among our species would this harvest be possible, for we alone make use of the digging stick. In addition to harvesting tubers, human children, like the offspring of other primates, participate very actively in foraging. A child of five can supply herself with up to half the calories she consumes daily, not independently of course, but by imitating her mother and following her instructions. To complete our description of the Hadza diet, we must also mention wild honey, another important food that is gathered by both sexes without distinction.

Kristen Hawkes, James O'Connell, and Nicholas Blurton Jones first described their "grandmother hypothesis" in the American journal *Current Anthropology*. This scientific publication has the admirable practice of providing space after its feature articles for the reaction of other authors, which encourages reflection and adds perspective to the new ideas presented. The grandmother hypothesis stimulated various reactions. In the first place, some said, longevity among primates corresponds so directly to the duration of development that it is difficult to believe that it would have increased in order to produce the phenomenon of menopause. It seems more logical to think that we live longer than chimpanzees for the same reason that it takes us longer to reach adulthood, which has something to do with our large brains.

We would also want to know whether the loss of fertility on the part of mature women drastically reduced the hopes of mature men for more children. In other words, in the hunter-gatherer societies studied by anthropologists, how often do older men have children with young women, i.e. women of the following generation? Is it possible that men also generally cease to procreate when their mates reach the age of menopause? In this case there would be a theoretical distinction, but no practical difference, between grandmothers who are physiologically post-menopausal and grandfathers who are also effectively post-reproductive. But among the Hadza it is unimaginable that a grandfather would lend a helping hand exclusively to his daughters and grandchildren, because the Hadza man's contribution is embodied in the hunt, and when meat is obtained it is shared with the entire group. Fathers and grandfathers feed not just their own families, but the group as a whole.

What's more, the grandmother hypothesis could only be valid if daughters lived with or alongside their mothers after reaching adulthood, rather than emigrating from the group. Chimpanzees follow a different pattern. Female chimps emigrate and lose all contact with their mothers after reaching sexual maturity. Among gorillas, both sexes emigrate. Neither sex leaves the group among the orangutangs and gibbons, but they *have* no social groups as such; orangutangs are solitary, and gibbons live in couples. All of this tells us that among the primate species most closely related to ours, there is no case of a matrilocal society, in which daughters stay in their birth group after reaching sexual maturity while males emigrate out of theirs. What is more, the majority of hunter-gatherer groups on record were or are patrilocal, meaning that sons stay in their birth groups while daughters leave theirs. Based on this evidence, many researchers believe it more likely that prehistoric hominids were also patrilocal, although Hawkes and her colleagues disagree.

There is another argument that should be considered in addition to the reactions to the proposed grandmother hypothesis that appeared in *Current Anthropology*. The hypothesis rests largely on the existence of an important vegetable resource that children cannot exploit unaided, the ekwa tuber. But the ekwa is toxic in its natural state, and must be roasted in order to be made edible. The problem is that we are not at all sure that fire was used in any systematic way more than 200,000 years ago. The evidence for earlier such use is shaky at best and seems to indicate that it was employed only sporadically.

After a great deal of speculation, it seems that we are going to remain in ignorance over the origin of menopause, in other words, the question of why the length of women's fertile period remained constant even though their life span increased. The explanation provided by the grandmother hypothesis has too many weaknesses to elicit the "Aha!" that fully convincing explanations spontaneously produce. Instead it suggests new questions. The crux of the problem may be that the hypothesis attempts to analyze natural selection on the basis of a comparison between individuals. It asserts that women who invest their time and energy in the feeding of grandchildren will, in the long run, have more descendants passing on their genes than women who do not provide food to their grandchildren, instead continuing to bear children of their own. I don't think that a satisfactory solution to the problem will be found by pursuing this comparison of individual outcomes. I would employ a theoretical framework that assumes a higher level of natural selection, one based on competing human groups. Ignacio Martínez and I wrote a book called *La Especie Elegida* (*The Chosen Species*) in which we made an effort to follow this line of reasoning. We proposed rational explanations for the social and cooperative behaviors of individuals such as grandmothers who dig tubers for the exclusive consumption of certain of their grandchildren, or fathers and grandfathers who hunt for the benefit of their entire group.

Finally, the central question we have to ask with regard to the hypothesis is to what extent the Hadza can be considered a universal model for an explanation of human evolution. The relative importance of vegetable foods in the diet, for example, varies dramatically among peoples and regions. The caloric intake of the Ache people of Paraguay is primarily of animal origin, and Inuit dependence on meat is even greater. Among these groups, recently weaned children unable to fend for themselves are often provided with food that had not long before brandished horns and pranced on hooves! Kaj Birket-Smith, a scholar well known for his work on the Inuit, commented in 1927 that the quantity of carbohydrates in the Eskimo, or Inuit, diet was minimal compared to that of animal fats and proteins. To partially compensate for this imbalance, whale liver and the stomach contents of reindeer were treasured delicacies, the former because it was rich in the carbohydrate glycogen, and the latter because it was a mass of fermented vegetables.

My conclusion is that human hunter-gatherers are very adaptable. The question is when this ecological flexibility developed. I think that it was two million years ago, and that it made the human migration from Africa possible. But the Iberian Peninsula is situated at an intermediate latitude between the equator of the Hadza and the far north of the Inuit. We need to determine the nature of the economic model that we can confidently attribute to the prehistoric inhabitants of the peninsula, and what sex-specific roles it entailed.

Foraging

To address this question, let us think about the possibilities for gathering plant food on the peninsula. At first, one might think that the peninsula would be a very problematic place to forage for vegetables. Otherwise, wouldn't there be a lot of monkeys living in Europe? We already know that the only primate to share the continent with us over the last million years has been the barbary macaque. But let me describe the possibilities available to a mammal that today occupies an ecological niche similar to that posited for prehistoric humans. I am referring to the brown bear, who lives from hunting, from carrion, and from foraging. The last Iberian brown bears live in the Cantabrian Range, primarily in Asturias and Castilla-León, and to some extent in Cantabria and Galicia. They also inhabit both slopes of the Pyrenees. The Pyrenean population is greatly reduced and is facing extinction. No more than eight individuals now survive, and although attempts are being made to increase their number by transporting bears from Central Europe, prospects for success are not great. I would be deeply saddened but not at all surprised to see the Pyrenean bear follow the bucardo, or Pyrenean mountain goat, to complete extinction. The Cantabrian bear population is also small. About sixty to eighty individuals survive, but unfortunately, they are divided into two nuclei, or isolated populations. The eastern nucleus enjoys more favorable conditions, living in the protected areas of Saja, Fuentes Carrionas, and Riaño. The western nucleus lives primarily in Somiedo and the Ancares range.

Since today's bears all inhabit the Euro-Siberian floral region of Spain, the model that I am going to explore applies only to this environment. Sad to say, our once plentiful Mediterranean bears are but a memory. In the first half of the fourteenth century, King Alfonso XI of Castilla y León ordered that a book be written describing large game hunting in his realm. The book, appropriately entitled *el Libro de la Montería* (*The Book of Hunting*) attests to the prevalence of bear throughout the kingdom, as far south as Tarifa and Algeciras on the Strait of Gibraltar, and describes the most promising places to hunt them. Madrid's coat of arms portrays a bear; the kings of old hunted them at the

Monte del Pardo, an immense oak forest just outside the capital. According to the royal chronicles, Philip II killed two in the sixteenth century. The bears "were doing a lot of damage to that land," the account says, so the man known in Spain as "Philip the Prudent" dispatched one with a crossbow and the other with a harquebus.

In the paragraphs that follow, I will describe the bears' diet based on the fieldwork of Rafael Notario, Gerardo Causimont, and Roberto Hartasánchez. I will intermingle information from the Pyrenees and the Cantabrian Range, since our primary interest is in what the ecosystems of the past may have had to offer. I will describe the bears' movements and activities over the course of a year, beginning with the spring.

In spring, the bear has just emerged from its winter abode and from hibernation. If it is an adult female, she has given birth to one or two cubs. One big difference between bears and humans is that the bear is a solitary animal, while our ancestors hunted and gathered in groups. When the bear leaves its winter shelter in April, it has consumed its reserves of body fat and is hungry. In the case of a female with one or two new cubs, she must nourish herself and also produce milk for her young. But there is little wild food to be had in the spring, and the bear must cover a lot of ground in order to meet its nutritional needs. In the beech woods of the Pyrenees, the bear will feed on beechnuts and woodrush, a reed that grows in damp environments. In the Cantabrian oak forests, it will eat the acorns that were left uneaten in the fall and spent the winter under a blanket of snow. When the snow melts, the bears may find and eat the newly thawed bodies of animals who died and froze during the winter, perhaps from hunger or avalanches.

From late May until July, the bear will dig in mountain meadows for the hazelnut-sized and very nutritious tubers of the pignut, a white-flowered member of the parsley family. If it finds supplies of pignuts set aside by mountain voles, it will also eat them. The bear digs for the underground storage organs of many other plants as well, the bulbs of "bear garlic" among them, as well as various other roots, tubers, and buried stalks. It will continue eating shoots and leaf buds, as it did earlier in the spring, and will graze on tender grasses. Given the opportunity, it will kill and eat herbivores, whether wild or domestic. Cherries are an attractive treat in the early summer, and bears will often harvest them by climbing a cherry tree to snap off a limb heavy with fruit, returning to the ground to consume the cherries.

Vegetable food is scarce until August, when numerous fleshy fruits begin to mature. These are very rich in natural sugars and will be truly abundant by the end of the summer and the early days of autumn. They include wild pears and apples, several varieties of serviceberries, the fruit of the hawthorn and the holly tree, currants, blackberries, juniper berries, raspberries, wild strawberries, rosehips, sloe plums, blueberries, barberries, bearberries (naturally), and more. The strawberry tree, or madrone, will complement this long list when it produces its fruit in the late fall and early winter. Although most of these fruits are small in relation to the size of a bear, the quantity of fruit produced, in terms of pounds per acre, is surprisingly large. In a good year, an acre of wild blueberries (an acre is almost exactly three quarters the size of an American football field, including the two end zones) may produce nearly 200 pounds of fruit. And bears are just crazy about blueberries!

During the fall, the bear must acquire and store sufficient energy reserves in the form of fat to doze through part of the winter in a state of hibernation, or semi-hibernation actually, as I will explain below. Nuts, rich in oils and starches, are very important to this plantigrade's autumn diet. Hazelnuts, beechnuts, chestnuts, walnuts, and especially acorns meet this need. The bears will retreat to their sleeping places in December and January.

In addition to foraging for vegetable food, bears avidly seek honey in wasps' nests and either wild or domestic beehives. They turn over stones to find ants and their eggs, and they eat the larvae of wood-eating insects that they find in rotting tree trunks. When they are hungry they may peel the outer bark off of trees to get to the sweet, fleshy sapwood, or phloem, found underneath. They also eat the many mushrooms that pop up in the course of the seasons, using their keen sense of smell to hunt truffles in the fall and winter.

The food list above should make it clear that there is little to be gathered until late summer and that autumn is the most bountiful season. Food is plentiful for only four to five months of the year. The fact that our bears are obliged to hibernate from December to April, like the dormouse, the hedgehog, and the marmot, is testimony to the scarcity of food. All of these animals are mammals, which means that they maintain a constant body temperature. But the ambient temperature often drops below freezing during the winter, so their bodies would require an unusually high expenditure of energy in order to maintain that constant temperature as well as a normal level of activity. Since no food is available at this time of year to provide the necessary calories, the bear dozes lightly in what has been called a "winter lethargy," but it does not fall into a deep sleep like smaller hibernating mammals. The bears' body temperature drops five to nine degrees Fahrenheit, and their pulse and respiration rates slow slightly. But compare this with the change in the body temperature of common hedgehogs. As they sleep in their burrows, the hedgehogs' body temperature falls in proportion to the ambient temperature, reaching as low as thirty-nine degrees Fahrenheit.

Meanwhile their pulse rate drops to twenty beats per minute and their respiration rate drops to ten breaths per minute. A bear's survival of winter lethargy depends on his or her physiological "larder," the amount of fat accumulated during the previous feeding season. If it was a poor autumn for oleaginous fruit production, the bear may not live to enjoy the tender shoots of spring. Nature is cruel, or as my university ecology professor used to say, "Wherever there's a lot of life, there's a lot of death."

The situation is not very different in our Mediterranean forests. In the southern half of the peninsula, there are a few more tree species that bear fruit edible to humans, like the piñon pine and the hackberry, but the currant bush, the cherry tree, the sloe plum, the service tree, the hazelnut tree, blueberry bushes, and other fruit bearers are rare or completely absent.

I should point out that some of these trees, specifically the chestnut, the cherry, the walnut, and the piñon pine may not be native to the peninsula. They were introduced within historical memory, from the time of Roman colonization on, for the economic value of their fruit. Although it had been thought that prehistoric Iberians never knew these plants, there is a fossil record of chestnut, piñon pine, and walnut that predates the last glacial maximum. Possibly these trees were wiped out by the rigors of the coldest period and were later reintroduced by humans, but it is equally possible that they survived in a few relatively protected environments and later recolonized the peninsula, albeit with some human help.

If naturally occurring vegetable food is so scarce on the peninsula for much of the year during an interglacial epoch like ours, imagine how hard it must have been to survive by foraging during the glaciations, which reached intensely cold extremes every 100,000 years. The most recent glaciation, the one we are most familiar with, was particularly harsh. The pollen records studied by María Fernández Sánchez Goñi in the caves of the Cantabrian Range attest to that. Although these sites are no more than 1,300 feet above sea level, they contain no tree pollens. All evidence seems to indicate that the landscape outside the caves was very open. No doubt there were small wooded areas in especially well-protected zones or near the sea. Fossils of forest dwelling herbivores like red deer and especially roe deer and wild boar are often found in these sites along with reindeer fossils. But the wooded patches must have been very limited in size and distribution.

In the cave at La Carihuela, twenty-eight miles from the southern city of Granada and situated at 3,350 feet above sea level, José Carrión and other paleobotanists have demonstrated that the coldest and/or driest periods of the last glaciation produced a treeless steppe environment. And human beings do not hibernate...

It should be clear from this discussion that access to quantities of animal fats and proteins has always been a necessary resource for human survival in Europe, above all during its coldest periods. Vegetable foods must also have played an important role, especially in the late summer and fall, and particularly during the less frigid periods, when forests dominated the landscape.

Unfortunately, paleodietary research by means of stable isotope analysis requires the destruction of a small portion of the fossil being studied. To further complicate matters, when we are fortunate enough to encounter human fossil remains, they are generally very few in number, often pertaining to a single individual. But the diet of one individual does not necessarily reflect that of an entire population. Using the extensive human remains found in the Sima de los Huesos at Atapuerca, Alejandro Pérez-Pérez, an anthropologist at the University of Barcelona, has employed a less destructive research method. Alejandro has extensive experience in microwear analysis of the striations found on fossil tooth enamel. Using an electron microscope, he compares the tiny grooves found on fossil teeth with those on the teeth of modern populations with known diets. The results of this research have led him to conclude that the humans of Sima de los Huesos consumed very abrasive vegetable foods such as seeds, roots, and tubers. Vegetable foods are abrasive when they contain a lot of silica or when earth is mixed in with them. So we conclude that the foods eaten at Sima de los Huesos were tough, that they were not softened before eating, and that they were not very clean. Even with the naked eye, we can see that the teeth of the Sima de los Huesos humans wore out very quickly, and we know that a meat diet causes little dental wear.

Alejandro Pérez-Pérez tells us that vegetable foods were very important in the diet at Atapuerca, but his research does not tell us exactly what those foods were. We may find the answer in a classical text. In *Los Pueblos del Norte* (*People of the North*) by Julio Caro Baroja, I found a quote from the work of the Roman naturalist Pliny the Elder. It is taken from Pliny's *Natural History*, completed in the year 77 A.D. *Natural History*, Book Sixteen, paragraph fifteen: "*Glande opes nunc quoque multarum gentium etiam pace gaudentium constant. Nec non et inopia frugum arefactis emolitur farina, spissaturque in panis usum: quin et hodieque per Hispanias secundis mensis glans inseritur.*" I won't make any false claims about my knowledge of Latin, but perhaps a few words will jump out at the reader as they did for me. I will tell you that the first word, "*Glande*," means acorn. But from there the erudite reader may pick out *emolitur, farina, panis,* and *Hispania*, which produce a string something like "soften, flour, bread, and Iberia." Bingo! The English translation of the text is as follows: "It is certain that even today the acorn constitutes a treasure for many of the peoples, even in times of peace. Since there is a dearth of cereal grains, the acorns are dried, they are peeled, and the flour is formed into a loaf. Today, the acorn is employed in the preparation of desserts, even in Hispania." Pliny later comments: "It is sweeter toasted in the ashes." Julio Caro Baroja included this quote from Pliny in his book, as well as a quote from Strabo's *Geography*, Book Three, stanza 155, also referring to the peoples of northern Iberia, the Galicians, the Asturians, the Cantabrians, the Vascons, and the Pyreneans. "All these mountain dwellers are sober; they drink only water, they sleep on the ground, and they wear their hair long in the feminine style, although for battle they tie a band around their foreheads... For three quarters of the year the mountain people eat only acorns. They let them dry, crush them, and later mash them, using them to make a long-lasting bread.... That is the way of life, as I have said, of the mountain dwellers." I am confident that the reader will find it in his or her heart to forgive me for not transcribing Strabo in the original Greek.

Gaius Plinius Secundus, Pliny the Elder, lived in the first century A.D., between the years 23 and 79. He represented the Roman emperor in Hispania as procurator and later had the misfortune to die in the line of duty while observing the famous eruption of Mount Vesuvius. Perhaps his scientific curiosity was excessive. He was overcome by the volcano's gaseous fumes. Strabo was a Greek geographer who was born in 63 or 64 B.C. and died about 23 A.D.

Based upon these classical sources, I think it reasonable to conclude that acorns can feed a considerable number of people during the period of their abundance in the autumn, as long as those people are "sober," as Strabo put it. If prehistoric humans knew how to dry acorns, crush them, make bread with them and then store it for subsequent use, then this resource could have helped them survive for an even longer part of the year. That process does not seem very difficult, but we have no evidence that they employed it. Nevertheless, not even acorn bread and other vegetable foods can sustain humans throughout the year in ecosystems as markedly seasonal as those of Europe and much of Asia, not even in a warm period like the present, much less during glaciations. Meat and animal fats were a necessary resource for human survival at our latitude and above.

For this reason, it seems appropriate to turn our attention to other animals more carnivorous than bears in order to establish comparisons with the probable human experience. The best comparison we can make on the peninsula is with the wolves. In fact, the wolf populations north of the River Duero occupy the same ecosystems, described above, as the Iberian bears. Wolves hunt the horse, the goat, the sheep, and cattle, the domesticated descendants of species; and they hunt the feral ungulates of the region: the red deer, the roe deer, the chamois, and wild boar.

There is another source of animal-based food that is actually gathered rather than hunted. I'm referring to the collection of mollusks, crustaceans, and fish from rivers, estuaries, and tidal pools. Even today many plants are used surreptitiously to poison rivers and kill fish, plants such as common mullein, oleander, water hemlock, cannabis, giant fennel, and spurge flax. Prehistoric humans may also have known of and used these plants, and they may have harpooned fish or caught them with their hands, at least occasionally. But fishing does not appear to have become economically significant until the Upper Paleolithic, the time of the Cro-Magnons, when the human diet apparently diversified to include just about every possible animal food, from rabbits to mollusks, and possibly a wider range of vegetable foods than in previous times as well. In the final stage of Magdalenian industry towards the end of the Upper Paleolithic, people fashioned some refined and beautiful harpoons, with one or two rows of fine teeth, out of deer antler. These harpoons were surely used for fishing. It seems that fresh-water fishing had become increasingly important at some European sites, especially the fishing of salmon, which seasonally abandon the sea and ascend the rivers.

While there is limited evidence for the consumption of seafood before the days of the Cro-Magnons, the shells of sea mollusks do not begin to appear with any abundance in the context of human activity until the Upper Paleolithic. We have to remember that at that time the coastline was far from its present position, and the rising sea level in the Holocene surely submerged most coastal evidence. By the Mesolithic period of our Holocene Epoch, Europe was completely free of ice. Human populations accumulated huge quantities of shells in many coastal areas, leaving them heaped together in large piles, indicating that mollusks were an intensively exploited resource. People consumed both sea creatures that cling tightly to rocks, such as periwinkles, limpets, and mussels, and those that burrow into sandy underwater environments, such as clams. The remains of sea urchins, crustaceans, and fish have also been identified among the accumulated shells.

Mesolithic humans lived in the north of the peninsula, on the border between today's Asturias and Cantabria, and practiced an economy largely based on seafood. Their lithic technology was characterized by crudely fashioned blades that they used to separate mollusks from the stones they clung to and to open shells. The most typical of these blades ends in a point and has been called an "Asturian pick." Large accumulations of discarded shells are also found on the banks of the River Muge, a tributary of the Tajo, upstream from Lisbon. In the valley of the Sado, south of Lisbon, there are large accumulations of shells, dating from the same period about 7,000 years ago. The coastal people of the Mesolithic fished with hooks and nets. Since we occasionally find the remains of deep-sea fish in their shell piles, one may imagine that they also ventured some distance from the coast in small vessels.

Hunters or Scavengers?

We have seen that prehistoric humans at higher latitudes (far from the equator) would necessarily have had to supplement a vegetable-based diet with meat in order to obtain an adequate supply of calories. But the ecological niche occupied by a hunter is quite distinct from that of a scavenger. While all carnivores dabble in both activities to some extent, it is worthwhile to investigate whether prehistoric Europeans belonged primarily to the lion clan or the hyena clan. Fossil remains of herbivores are often encountered in caves like those at Atapuerca. But these animals did not graze inside caves. Their bodies must have been brought there by humans or carnivorous animals. We need to find out which.

If humans were the exclusive authors of this activity, then the herbivores' bones would in every case show signs of having been butchered. Prehistoric humans used the cutting edges of their stone tools to skin and dismember carcasses, to cut tendons, and to separate muscle from bone, leaving clear and characteristic marks that have been studied extensively and in detail by modern specialists. Fossil bones may also have natural striations, but if cut marks appear at strategic sites and in set patterns indicating the removal of skin, the dismembering of the animal, or the cutting away of meat, then there is no doubt of human agency.

Carnivorous animals, in contrast, leave tooth marks on the bones of their prey. Humans characteristically open bones along the shaft to extract their marrow, a technique very different from that of carnivores, who often bite open the upper ends of the humerus and the femur where they articulate with the shoulder and hip, respectively. To complicate matters, hyenas also open bones along the shaft to extract marrow, often creating problems of interpretation.

Finally, even if we know that an herbivore was brought to a cave and consumed there by humans, we are still not sure if the animal was hunted or obtained as carrion. Either a human or animal hunter, if there was one, would generally have had first access to the meat, but humans and carnivores also keep an eye out for animals that may have died accidentally or from some other natural cause. To address this problem, we can consider the array of bones found at the site. If they form a complete skeleton, we can presume that the humans were hunters and that they obtained the entire carcass and carried it to their living space, either whole or cut in pieces if it was inconveniently large. When the bones that hold the greater part of the meat are missing, such as the hips, the femur, the tibia, the humerus, or the scapula, we have to suspect that the humans arrived late to the feast and had to make do with leftovers. Most of the cuts of beef that we eat come from favored parts of the anatomy. Chuck or shoulder come from the front quarter; rump, round, and shank come from the back quarter; and loin comes from the area around the spinal column. If we find only the heads and forelimbs of an ungulate in an archeological dig, we suppose that the humans had only secondary access to the carcass, and these less desirable animal parts were retrieved as carrion.

Experts in this field examine herbivore fossils and compile their statistics carefully, employing a lot of patience and a good dose of common sense. They also consider other data, for example the age of the herbivores at death. If there is an abundance of very young animals, the cave was likely occupied by humans in the spring and summer, shortly after the prey animals were born. In any case, humans always carried their food home to the cave to prepare and consume it as they pleased. If we want to explore the site of the hunt itself, we look in the surrounding area. So many sites are known, and they differ so much from each other, that we had best concentrate on a few of the exceptionally revealing ones.

The site at Boxgrove burst into the public consciousness in the spring of 1994, when the discovery of a human tibia there was announced in the journal *Nature*. At that time, the Boxgrove tibia was, along with the Mauer mandible, the oldest human fossil in Europe, some 500,000 years old. A few months later, in the summer of the same year, we were finding human remains 300,000 years older in the Gran Dolina at the Sierra de Atapuerca. Although Boxgrove enjoyed no more than its allotted fifteen minutes of fame in the mass media that spring, it had already been an important prehistoric site for several years. It is a magnificent site from which fossils of many species have been retrieved, together with numerous flint utensils, mostly oval-shaped hand axes. It has also yielded deer and megaceros femurs and antlers modified for use as soft hammers. These soft hammers, as opposed to hard hammers of stone, were used to add the delicate final touches in sophisticated tool fabrication.

Boxgrove is in West Sussex, seven and a half miles from the southern coast of England and the Channel. Half a million years ago, it was a lagoon frequented by humans and animals, on a broad coastal plain that stretched from the seashore to a series of white cliffs. It is an open air site with a number of features that make it unusually advantageous for the study of prehistoric human activities. Covered by the placid waters of the shallow lagoon, its numerous fossils and utensils long lay undisturbed, practically just as they had been abandoned.

Half a million years ago, humans at Boxgrove fashioned tools and butchered and consumed large herbivores like megaceros, deer, bison, and rhinoceros. We do not know for sure if these animals were hunted by the humans or by other predators, perhaps the wolves and bears whose remains have also been found there. Mark Roberts, the archeologist who directs the Boxgrove excavation, is convinced that humans killed most of the herbivores by means of cooperative hunting. The scientific team has encountered the scapula of a horse that had been punctured by the sharp point of a wooden projectile. The projectile itself has not been found and probably never will be, since wood very rarely fossilizes. A third possibility is that at least some of the animals died on their own, without being brought down by any other creature, that is. There is no way to determine how long a period this site records, but it is undoubtedly more than just a snapshot in time. This is true for the majority of prehistoric sites. Typically, many layers of events are superimposed upon each other, though they took place over long periods of time within the immensity that is the past.

Roberts is certain, however, that humans usually got to the meat before the carnivores did. The carnivores' tooth marks are superimposed over marks previously produced by the humans' biface cutting tools. The fact that humans got first dibs on the meat could mean that they found the animal dead of natural causes, or that they killed it, or that they snatched it from a carnivore before it could enjoy the reward of the hunt. The oft-mentioned Hadza steal meat from predators just as much as they hunt for themselves. Their search for carrion is not systematic, nor does it limit their other hunting and gathering activities. They are simply opportunists. They closely observe vultures circling overhead and the cries of lions and hyenas, they locate the carrion, and they retrieve it.

In 1988, James O'Connell and Kristen Hawkes spoke at El Escorial, an enormous sixteenth-century palace, library, monastery, and mausoleum near Madrid, as part of a summer course that I had organized with Leslie Aiello. They reported that when a group of Hadza arrive at the scene of fallen prey, hyenas, leopards, and even lions retreat and leave everything to the humans. "And if they don't?" I asked. "They are run through with arrows," was the response. The Boxgrove humans did not have arrows, so they would have had to advance farther towards the predators. I will return to this question. Suffice it to say that Mark Roberts believes that the humans themselves were the hunters who brought down the herbivores that they later consumed at Boxgrove.

Levels F and G of the Arago cave near Autavel in the French Pyrenees are nearly as old as the site at Boxgrove. The famous archeologist Henry de Lumley excavated them many years ago. A significant number of mouflon sheep (*Ovis antiqua*) have been found at these two levels, eighty-three of them at level F and forty-two of them at level G. According to Hervé Monchot, they were hunted by humans and transported whole to the cave for consumption. The majority of the mouflons were young adults, often weighing more than 225 pounds. Hervé Monchot concludes that these humans of Arago cave were hunters, not carrion eaters, and that they favored young adults rather than the immature and overaged individuals preferred by wolves, felines, and hyenas. The hunting practices of wolves in the Sierra de Culebra in Zamora confirm that red deer of very tender age and of advanced age are at much greater at risk of being attacked and killed than young adults, so much so that only between thirty and forty-five percent of fawns reach the age of six months.

The Schöningen Spears

In January 1977 I traveled to the German city of Jena with Eudald Carbonell and Jan van der Made, at the invitation of our colleague Dietrich Mania. Many important discoveries were being made at the famous site of Bilzingsleben, and we wanted to see the place with our own eyes. While in Germany, we also wanted to visit Hartmut Thieme, whom I had met the previous summer in Burgos, when he had presented some remarkable findings at a symposium on the first Europeans. Hartmut picked us up at the train station and drove us to the site. It is called Schöningen and it is some sixty miles east of Hannover. It was snowing that unusually cold January, and Hartmut's car skidded along a frozen highway through a world of pure white. When we arrived at the site, we found ourselves in front of an enormous open pit where a giant mechanical shovel worked day and night, irregardless of cold or snow. It was a coal operation, a surface mine. The huge machine advanced along a long front, devouring everything in its path and leaving only a gaping cavity in its place. The material extracted included a seam of coal. This was separated from the earth and associated detritus, which were later deposited behind the advancing behemoth and replanted with trees.

A team of researchers led by Hartmut had spent years removing everything of archeological significance from the path of the shovel. But the team had most recently made some discoveries of such overwhelming importance that the mining machine had been obliged to take a detour of several years' duration while research could be conducted.

We entered the excavation through a domed plastic tunnel. Despite the frigid air outside, the archeological team was working in tee shirts, thanks to a powerful heating system. I will never forget what I saw next. A fossilized horse pelvis lay on a black bed of peat . . . with a 400,000-year-old wooden lance of the same age sticking out of it! Hartmut smiled at our exclamations of wonder. He was well aware that he had made an historic discovery.

Hartmut Thieme has found four well-preserved lances in Schöningen so far. One measures six feet in length, another measures seven feet 4.5 inches, and the third measures seven feet 6.5 inches. The one we were looking at that day is broken into four pieces and also measures over 6.5 feet in length. These lances were fashioned from the trunks (not limbs) of young Norway spruce trees. Norway spruce is a conifer that does not occur naturally on the Iberian peninsula, but is frequently found planted in parks and gardens. They are also sold as Christmas trees. They look much like firs, but their cones hang from the limb rather than stand erect on top of it. The growth rings on the Norway spruce of Schöningen were very close together, indicating slow growth due to a cold environment. An analysis of fossil pollen at the site suggests a landscape of meadows with some pines, Norway spruce, and birches scattered about.

I have been referring to these wooden weapons as lances, but there is some question whether they were actually used as a lance: held firmly at one end and thrust forward to stab or pierce. They may have been designed instead as spears to be thrown, more like a javelin than a pike. Based on their engineering, Hartmut Thieme believes that they were throwing weapons. A lot of skill went into their production and into ensuring a forward center of gravity, close to the point. This would have served to maximize their range in flight. They weighed about 4.5 pounds, not too heavy for the muscular Europeans of the time. With these weapons, people would have been able to kill their prey from some distance, avoiding the danger of being gored or kicked.

Horse remains are plentiful at Schöningen. They show signs of having been quartered and having had the meat stripped from their bones. Four hundred thousand years ago, groups of human hunters must have lain in wait for herds of the animals on the shores of a lake, perhaps under the cover of an early morning mist. They would have drawn near to their prey, creeping through tall rushes, until they were in throwing range, whereupon they would have launched a flurry of spears. If they were in luck, an animal would have been run through and fall dead. Every horse they brought down would provide the human group with hundreds of pounds of the meat that they so badly needed in order to survive in that cold environment.

Elephant Hunting on the High Plains

Since we are discussing big game hunting, why not bring in the elephant hunt? The elephant is the largest possible game animal on the face of the earth. For many years, the towns of Torralba del Moral and Ambrona, in the province of Soria, have occupied a place of prominence in the prehistory taught at universities around the world. Numerous elephant fossils have been found in association with biface and other stone instruments near these towns in the valley of the River Ambrona, also known as the Mansegal, but in either case a tributary of the Jalón. Many researchers have concluded that elephants were hunted by humans some time after the days of the Schöningen spears. Other scientists believe that the finds have been misinterpreted and that the elephant hunts, about which so much has been said, never really took place.

In 1888, a trunk line was being constructed to link Soria to the existing Madrid-Zaragoza railroad. It had been decided that the junction would be at a new station to be constructed in the town of Torralba del Moral. When construction on the station began, some giant bones turned up and stimulated considerable interest. From 1909 to 1911, the first scientific excavation to explore the phenomenon was led by Enrique Aguilera y Gamboa, the Marqués de Cerralbo, who was an aristocrat with an interest in archeology, and a priest named Justo Juberías. The Marqués announced their finds at the 1912 International Congress of Anthropology and Archeology in Geneva. Excavations were renewed between 1961 and 1963 by the famous American paleoanthropologist, my good friend F. Clark Howell. He and L. Freeman also began another round of study at the site in the early 1980s.

The horse and the straight-tusked elephant *Palaeloxodon antiquus* are the most common animal remains retrieved at these sites. Horses clearly prefer to feed in extensive herbaceous environments, i.e. grassy steppes. Elephants grazed on these plains too, but they are browsers, so their diet would also have included tree foliage, fruit, bark, and twigs, as it does today. Another member of the same ecosystem, although less frequently found in these sites, was the steppe rhinoceros, *Dicerorhinus hemitoechus*. Red deer, fallow deer, and aurochs have also been found. Carnivores, on the other hand, have rarely turned up, being limited to an occasional hyena, fox, wolf, or lion. Although the local vegetation has been described as alpine, straight-tusked elephants and steppe rhinoceri are not compatible with extreme cold, so the climate at the time in question must not have been very different from today's. It may have been an interglacial or perhaps an interstadial, one of the warmer intervals within a glacial period. In addition, there are signs of macaque, a Mediterranean primate who cannot survive intense cold, or rather cannot live without the vegetable foods that such cold precludes. On the other hand, it must be said that Clark Howell is not certain that macaques were indeed present.

Torralba and Ambrona are situated on a high plateau; the archeological sites are at about 3,600 feet. Animals would have climbed up there to graze when the vegetation in the lowlands withered. The plateau is well situated between the Duero, Tajo, and Ebro drainages. It would be worthwhile for a modern traveler on the Madrid-Zaragoza-Barcelona highway to stop at Medinacelli and visit this old village and its Roman arch, with a view of the strategic River Jalón, a tributary of the Ebro. Just a few miles farther on is the hamlet of Ambrona, and down the road to Torralba, on the left, lies the Loma de los Huesos, an archeological site on the plateau's edge that was prospected in 1911 by the Marqués de Cerralbo and later excavated by Clark Howell. The bones of various elephants discovered in 1963 are on display there, *in situ*, protected modestly but efficiently by four walls and a roof built by Emiliano Aguirre, our warmly appreciated director of excavations at Atapuerca. More fossils and stone utensils are displayed just a few steps away in a small orientation center. The view from the plateau is impressive for its austerity. There are no trees in sight, although this is the result of human intervention, not of climatic conditions. It is not difficult to imagine herds of elephants and ungulates grazing on the sea of ridges that extend into the distance.

Then as now, marshes and lakes dotted the landscape. It has been thought that humans may have frightened the elephants so as to drive them into the soggiest areas, where they would have been bogged down in the mud and made easier hunting. But why would a herd of massive elephants be so afraid of humans, even a group of shouting men, that they would lose their heads and stampede directly into a muddy trap? In Ambrona alone, Clark Howell has excavated the remains of forty-seven elephants, although he does not believe that they all died at the same time. It could be that the human hunters set fire to dry prairie grass to drive the beasts in the desired direction. A diorama in the National Archeological Museum in Madrid illustrates just such a scenario. Organized hunts of this nature carried out over a long period of time would have produced the large accumulations of fossil remains found at Torralba and Ambrona.

Another prominent researcher named Richard Klein proposes a diametrically opposed interpretation of the very same evidence. He believes that the widely divergent ages of death among the elephants suggests nothing more than their expected natural mortality, what is known as an attritional profile of remains. A catastrophic profile, on the other hand, could be explained by death as the result of an indiscriminate hunt. In sites with an attritional profile, older individuals predominate over young adults in the prime of their reproductive lives, just as they do in human cemeteries. In the case of sites with a catastrophic profile, the opposite is true; young adults predominate over older individuals at the site simply because older individuals are a minority within live populations.

Lewis Binford, another prominent archeologist, sees a closer association of stone instruments with horses, deer, and aurochs than with elephants. His new studies cast grave doubt on the scenario of large-scale elephant hunting in the Mansegal Valley and suggest that humans may have found and helped themselves to some elephant cadavers, but only as an unplanned, occasional, and sporadic activity. This image of humans as hungry opportunists is a far cry from that of proud organizers of mass elephant hunts. Clark Howell has responses for this and other criticisms of his hypothesis of organized elephant hunting, and the debate is far from over. Beyond the physical capacity of prehistoric humans to hunt elephants, the crux of the polemic is in their mental capacity to develop and execute complex hunting strategies based on seasonally predictable conditions. Planning is powerful evidence for consciousness.

But while the truth is elusive at Torralba and Ambrona, there is another site of approximately the same age where the record is quite clear. Áridos sits on a terrace alongside the River Jarama, very close to Madrid. The remains of two elephants have been found there, about 650 feet apart. One is a young female, at Áridos I, and the other an old male, at Áridos II. There are no marks of carnivore teeth on their bones. We do not know that the animals were necessarily hunted down by humans, but there is no question that humans had the first and only access to their meat. The prehistoric inhabitants of Madrid engaged in large-scale meat-cutting projects in the valley of the Jarama.

The archeologists digging at Áridos, directed by Manuel Santonja and Ángeles Querol, have determined that the butcher's tools used there were manufactured on-site. In many cases, it was possible to reconstruct the tool production sequence in reverse by fitting together an implement with its corresponding residual material. Together they would form the flint nucleus or the quartzite stone from which they had been separated long before. It was also observed that when the stone tools' cutting surfaces dulled, they were sharpened on-site in order to continue the butchering. Thus the complete sequence of events at Áridos has been reconstructed, from the time that humans arrived until the time they left. One interesting aspect of the process is that while the quartzite was obtained right on the banks of the Jarama, the flint had to be brought from Manzanares, some two miles away. This is unequivocal evidence that these people planned their activities. The retrieval of flint may have been a minor or short-term plan, and it may have occurred after the possibly accidental discovery of an available carcass, but it was planning nonetheless.

In this case, and unlike the description of most archeological sites provided some paragraphs above, Áridos I and II represent two isolated moments in geologic history, not the superimposition of many events taking place over a long period of time. This facilitates the analysis of the sites, especially if seen in contrast with the cases of Torralba and Ambrona, which doubtless represent the sum of many events and where, according to the critics of the organized-hunt hypothesis, geological phenomena have also intervened, modifying the fossil and artifact accumulations.

Humans were undoubtedly present at the Torralba and Ambrona sites at various points in time, but where in particular and under what circumstances? We cannot provide definitive answers to these questions, but Manuel Santoja and Alfredo Pérez-González are currently conducting excavations there that may tell us. Alfredo is also the lead geologist at the Atapuerca project. For now, the most widely accepted hypothesis is that the ancestors of the Neanderthals did not organize and carry out a slaughter of elephants at Torralba or Ambrona in the Iberian Middle Pleistocene, but that they did exploit occasional opportunities to carve meat from the cadavers of animals who had died of natural causes.

More researchers believe that humans did hunt woolly rhinoceros and woolly mammoth at La Cotte de Saint-Brelade, located on what is today the Channel island of Jersey, but which was attached to continental Europe at the time. Here the animals may have been stampeded off a cliff towards the end of the Middle Pleistocene, after the time of Torralba and Ambrona. By this time the humans in question would have been almost fully evolved Neanderthals.

If hunting pachyderms is dangerous, hunting large carnivores is no picnic either. Some 200,000 years ago, a large number of bones were deposited on the banks of a river in Biache-Saint-Vaast, near the Strait of Dover and not far from La Cotte de Saint-Brelade. They included bones of roe deer, red deer, megaceros, aurochs, Merck and steppe rhinoceri, horses, and smaller, now extinct equids. The environment was not terribly cold; it was an interstadial, a less extreme interval in the course of the penultimate glaciation. Red deer and aurochs grazed in forest clearings. Roe deer nibbled at leaves and tender buds.

Horses and rhinoceri pastured in broad meadows, and megaceros waded through swampy clearings, their wide antlers unencumbered by the obstacles of the nearby forest. All of these animals were eaten and possibly hunted by humans. Two crania with highly evolved Neanderthal characteristics have been found at the same site.

The surprising thing about Biache is that various brown bear and cave bear fossils have also been found, at least ten of them at a single level, base II, and that they had been skinned and butchered. Their meat had been cut from the bones, which had been broken open and the marrow removed. In sum, they had been thoroughly consumed. Patrick Auguste, the paleontologist who has studied these bear remains, believes that they were hunted by the forebears of the Neanderthals and not obtained as carrion. The evidence for this is that most of them were young adults, not the very young and very old individuals expected to die of natural causes.

Many other open-air sites from the Middle Pleistocene have been studied and could be cited in an examination of human-faunal interaction. Evidence at Bilzingsleben, for example, is interpreted to support the hypothesis of the human as a dweller in highly organized encampments and a consumer of large herbivores, engaged in hunting as a sophisticated social activity. The space available here to elaborate on this debate is limited, so I will state my conclusions as clearly as possible. Given the almost complete lack of vegetable food in the winter and spring, there is little doubt that animal fat and protein were necessary for human survival in Europe. Meat could have been obtained by hunting, by recovering carrion, or by any combination of the two. I do not think that carrion eating is an alternative to hunting or gathering in our case, nor do I think that primates are equipped to be "professional" carrion eaters, but rather occasional ones, as a complement to other activities. This is in keeping with the activities of the Hadza. The prospects for food gathering on our continent are seasonally determined, so the exploitation of carrion can only be complementary to the principal activity, i.e. the hunt, for much of the year.

Finally, I am absolutely convinced that the enormous physical strength of the humans in the European Middle Pleistocene, of which the fossils at Sima de los Huesos leave no doubt, was an adaptation to their need to kill animals of prey at short range. The idea strongly promoted by some authors, such as Lewis Binford, that these humans as well as the Neanderthals were weak and defenseless creatures is completely ridiculous to me. This hypothesis asserts that they were limited to vegetable collection and occasional leftovers abandoned by carnivores. In constant fear of the latter and dependent on the occasional lucky find of meat, they were the most pathetic mammals in the ecosystem. Despite such a miserable existence, they were the animal with the largest brain capacity and along with the elephant, they lived the longest...?

Instead, I envision a hunting party of powerful individuals, each weighing some 200 pounds or more of pure muscle, clad in bearskins and armed with long well-sharpened lances, a group from which lions would turn and flee.

The Neanderthals would appear at the end of this period and live similarly, retaining the physical strength of their forebears.

The first modern humans in Europe, the Aurignacians, were also very strong. But they were a distinct species, with narrower trunks and hips. The paleoanthropologist Steven Churchill has studied an Aurignacian humerus found at Vogelherd, a German site, and has described it as being as robust as a Neanderthal humerus, though it differed in other ways. But that much alone tells us that this was a very strong arm indeed. The Cro-Magnon skeleton would grow less robust throughout the Upper Paleolithic, and this trend would continue in the Mesolithic. The appearance of new and deadly hunting technology in the form of the bow and arrow and the spear thrower, or atlatl, may account for this diminished robustness.

The javelins of the Middle Pleistocene were generally sharpened at one end, but they may have occasionally been fitted with stone points instead. Three fragments of fir shafts, cleft at one end to accommodate stone points, have been found at one of the Schöningen sites. If they are what they seem to be, these would be the first known weapons to incorporate two materials, in this case stone and wood. The Neanderthals almost surely used the many stone points found in Mousterian sites for the same purpose. Long spear points of bone and antler first appear in Aurignacian industry. They were attached to wooden shafts, but the projectile was still launched with the unaided strength of the hunter's arm. They were what we call assegai points.

The spear launcher, or atlatl, is a short rod with a hook or groove at one end to hold the base of a long dart, while the other end of the launcher is held firmly in the operator's hand. The device effectively increases the reach and power of the throwing arm. The spear launchers that we have found were made of deer or reindeer antler or of ivory. They were often attractively decorated, indicating that they were objects that carried prestige. Most spear throwers, though, were made of wood, which is why they decomposed. Even today, hunting people manufacture wooden spear throwers to increase the thrust of their projectiles. It is thought that the device is an invention of the Solutrean technological complex, which followed the Aurignacian and the Gravettian about 20,000 years ago.

The invention of the arrow is a little harder to date, but some Solutrean points seem to have been designed as arrowheads, especially those with lateral fins or barbs and a central peduncle for fitting the head to the shaft. This construction is very typical of Solutrean technology in the region around Valencia. The oldest known arrow is 11,000 years old, about the same age as a figure engraved on a plaque at Grotte de Fadets in France that has been interpreted as depicting an archer. In eastern Spain, there are many cave paintings, collectively known as Levantine Art, that represent archers, but they seem to be less than 10,000 years old. I will discuss them further in the Epilogue.

The use of the spear launcher and the bow were revolutionary techniques that enabled humans to kill from a distance, manifestly altering the equilibrium between them and their prey. There is a tremendous difference between approaching a bison with a lance in hand and projecting a dart or arrow into it from a distance. We do not know if these people applied poison to their projectiles. If so, their effect would have been that much more terrible. Many authors believe that the disequilibrium brought about by human technology ultimately caused the extinction of many mammal species. If so, it would have been the first far-reaching ecological impact attributable to humankind, amply illustrating that such sinfulness is neither a symptom of modernity nor exclusive to industrial society.

The Last Mammoth

The woolly mammoth may be the creature most emblematic of the ice age, the Pleistocene. When the Pleistocene gave way to the Holocene, the woolly mammoth disappeared forever, along with the megaceros, the woolly rhinoceros, and the cave bear. At the same time, herbivore species that had long grazed together in Western Europe, like the reindeer, the musk ox, and the saiga antelope, retreated in different directions. While the reindeer and musk ox followed the retreating tundra northward, the saiga ante-lope followed the receding steppe to the east.

If these Eurasian transformations seem dramatic, the climatic catastrophe was substantially greater in the Americas, where it affected many more large mammal species. In North America alone, and counting only species weighing over ninety pounds, we can enumerate the following extinctions: Among the Proboscidea, the woolly mammoth and two other mammoth species disappeared, as did the mastodon, a distant but equally large relative. Many species of camels, llamas, moose and deer, and all the pronghorn antelopes (antilocaprids), save one species, perished. The musk ox and the flatheaded peccary, a relative of the pig, were lost. Among the cats, the large sabre-tooth genus *Smilodon* succumbed, as did the *Homotherium* discussed earlier. Cheetahs had survived in North America through the Pleistocene, although not the species we are familiar with today. The giant shortfaced bear, larger than any of today's species, also disappeared. The great extinction at the end of the Pleistocene wiped out many rodents, including a giant capybara and a giant beaver. The North American tapir was lost, as was the horse. The horse was reintroduced to the continent much later by Spanish conquistadors, and the descendants of runaway individuals formed the wild herds of the West, making possible the great horse cultures of the Plains Indians. The order Xenarthra, also known as Edentata, "toothless," is one of the old South American mammal groups that evolved on what was an island-continent in complete isolation from the remainder of mammal life for mil-lions of years. When the Isthmus of Panama was established and the Americas were joined, many "modern" mammals arrived, but the Edentata survived the crisis and even spread to North America. They were decimated by climate change at the end of the Pleistocene, and many species were extinguished, including the giant armadillo, the glyptodon, which was covered by a rigid bony shell like a turtle, and three families of terrestrial sloths: the megalonyx, the mylodons, and the megatheria, some of which reached enormous proportions.

The giant megatherium, one of these sloths, actually survived in South America until the Holocene. This species played a somewhat important role in the history of paleontology. The nearly complete skeleton of a large animal was found on the banks of the Luján River about forty miles from Buenos Aires and sent by the Viceroy to the Royal Natural History Collection in Madrid, arriving in September 1788. Juan Bautista Bru de Ramón, "animal preservationist and anatomical painter" at the Royal Collection, mounted it, studied it, and drew it, publishing a monograph in 1796 with five large etchings by the scientific illustrator Manuel Navarro. One of the plates, now very well known, depicts the skeleton mounted and standing on four limbs, while the others illustrated separate skeletal components. The great French paleontologist Georges Cuvier was drawn to the work and identified the animal as a large extinct Edentata, giving it the scientific name *Magatherium americanum*. Cuvier highly praised the work of the Spanish naturalist and declared it worthy of imitation. The skeleton can be seen today, just as it was mounted by Bru, in the National Museum of Natural Sciences in Madrid.

The discovery that large animals like the megatherium had existed in the past led Cuvier to formulate his theory that there had been a series of catastrophes throughout the earth's history that had repeatedly caused mass extinctions. Following each one of these catastrophes, God had engaged in a new act of creation, giving life to a new generation of living beings. The evolutionary theory of Darwin and Wallace later came to replace this approach, and today it is the only one that is accepted. Darwin himself was intrigued by the large extinct megatherium. In 1832, during his round-the-world trip on the *Beagle,* he wrote a letter from the River Plate commenting on megatheria fossils he had found and on the skeleton in Madrid.

Climate change may account for the widespread extinctions that accompanied the end of the ice age in Eurasia and the Americas, but some authors attribute them instead to the spread of our species to every corner of the planet, resulting in a wave of destruction that continues to this day. As of yet, there is no proof that humans have caused the disappearance of a single prehistorical floral or faunal species. In this discussion we should remember that the first species to feel the tremendous impact of our expansion were the other humans who inhabited Africa and Eurasia, *Homo erectus* and the Neanderthals. They suffered extinction thousands of years before the end of the Pleistocene.

We do know that none of the extinct species of the Americas had ever seen a human being there. Some of them were very large and slow, the giant sloths for example. One can only imagine that hunting them would have been child's play for the ancestors of the Amerindians. They could have used them for target practice. In other cases, there does not seem to be a clear or direct relationship between humans and the extinction of species in the Americas. Horses, for example, survived elsewhere. It is possible that at times humans altered the ecological equilibrium by exterminating some elements of the ecosystem. Eliminating the most vulnerable prey can cause a chain of extinctions that ultimately affects even the last links on the chain, the large predators.

If we are to consider the arrival of humans in the Americas as the cause of a widespread multispecies extinction, we are faced with a vexing circumstantial problem. The spread of humankind to the Americas and the climate change that signaled the end of the ice age were practically simultaneous, making it very difficult to determine the role of each of these factors in the rapid decline of biodiversity. The human species that populated the Americas is ours. No other humans, neither *Homo erectus*, the Neanderthals, nor any of their or our ancestors had ever before traveled so far. One reason for this is that probably no human species before ours had populated the Chukotskiy Peninsula of eastern Siberia. It is so cold at the Arctic Circle that survival there is very demanding. Then too, in warmer times, the Bering Strait could only be crossed by boat, but during cold periods when the sea level dropped and Alaska could be reached on foot, the passage was an appallingly inhospitable place to be.

The oldest evidence of a human presence in the Americas is archeological. Examples of a very beautiful stone instrument called the Clovis point have been found in many locations. Clovis points are fluted, from one to seven inches long, and beveled at the end that was inserted into a wooden shaft. They are very elaborately and carefully carved over the entire surface of their two faces. The settlements where the points have been found are no more than about 11,500 years old. This is the moment, during a glaciation, traditionally believed to have marked the arrival of humans on the American continent. Recently though, evidence of a human presence 1,000 years earlier has been discovered at Monte Verde in Chile.

The truth is that not a lot is known about how humans spread south through the Americas. Although part of Alaska was not helmeted with ice, human migration south would have been blocked by two other large ice barriers. The larger of the two, centered over Hudson Bay, covered all of Canada and extended south beyond the Great Lakes. It is known as the Laurentian ice sheet. The other ice sheet was smaller, covering the Coast Range in the Pacific west. At the glacial maximum 20,000 years ago, the two mantles merged, forming an impassable obstacle. But after the two icecaps again separated at a later and somewhat warmer date, humans probably found a narrow land corridor between them. On the other hand, they may have taken a sea route south along the Pacific coast to bypass the enormous ice mantle. In any case, pass it they did, and they quickly reached the Strait of Magellan, possibly extinguishing many species along the way.

But let's get back to the woolly mammoth, the species most emblematic of the ice age. It is believed to have disappeared 12,000 years ago in Europe, 11,000 years ago in North America, and 10,000 years ago in north-central Siberia, apparently its last refuge following the retreat of the glaciers. At one time, these dates were perfectly compatible with the hypothesis that human hunting was directly responsible for the disappearance of the mammoths, because 12,000 years ago humans had arrived in the extreme northeast of Siberia and had probably crossed into North America. But in 1993, three Russian scientists named S. L. Vartanyan, V. E. Garutt, and A. V. Sher published a surprising paper. They reported the discovery of woolly mammoth fossils between 4,000 and 7,000 years old on Wrangel Island in the Arctic Ocean, 125 miles northeast of Siberia's Chukotskiy Peninsula. This means that when the last remaining woolly mammoth perished, ancient Egyptians had already built the great pyramids! The forebears of these remnant mammoths had crossed from Siberia to what is now Wrangel Island when both places were part of Beringia. With the coming of the great Holocene thaw, the sea level rose and much of Beringia disappeared beneath the waves, creating the Bering Strait. The Wrangel Island mammoth population was left isolated and safe from human hunters on their island refuge. Another peculiarity of the Wrangel Island mammoths is that they evolved to be at least thirty percent smaller than their ancestors who had come from Asia and colonized the island. The average Wrangel mammoth weighed about 6.5 tons and stood eight to ten feet high at the withers.

Reduced size is not an unusual phenomenon among island populations. Evolution in conditions of isolation has produced other cases of dwarfism even more dramatic than that of the Wrangel Island mammoths. Dwarf elephants lived on many Mediterranean islands in the Upper Pleistocene, including Malta, Sardinia, Sicily, Cyprus, Crete, and various other Greek islands. The largest males among the *Palaeoloxodon falconeri*, who lived in Sicily, stood less than three feet high. For purposes of comparison, the largest known modern elephant was killed in Angola in 1955 and is on display at the Smithsonian Institution in Washington, D.C. Although African elephants rarely surpass 3.5 tons or 11.5 feet in height, this giant weighed eleven tons and stood thirteen feet high at the withers. The Asian elephant is a distinctly smaller species, similar in height but less stout than the woolly mammoth. Surprisingly, the Sicilian dwarf elephant evolved from the continental straight-tusked variety. Some researchers attribute its reduced size to a scarcity of food on the islands, while others point to the absence of large terrestrial predators. It may seem like a joke, but if a young dwarf elephant on the Mediterranean islands wanted to check for enemies, it would have had to look skyward, for the eagle was its only natural enemy. Actually the two hypotheses are related. Size itself is an advantage. Lions and tigers do not prey upon adult elephants, and sharks do not prey upon whales. But such intimidating size does not come cheap. The price is a life dedicated to consumption, assuming that sufficient food and water is even available. An adult elephant may eat about 650 pounds of food and drink more than forty gallons of water a day. In the absence of threatening predators, they could afford to reduce their size, thereby increasing the likelihood of filling their bellies.

The Russian authors mentioned above are proponents of the theory that the impact of climate change on the plant life of the steppe-tundra environment indirectly caused the mammoths'extinction at the end of the Pleistocene. They believe that the mammoths on Wrangel Island outlived their continental relatives because the habitat, to which they were adapted, managed to persevere on the island. Modern elephants graze and browse on larger vegetation, while woolly mammoths lived exclusively by grazing on arctic steppes. Climate change eventually left them hungry, even on Wrangel Island, but according to Vartayan and his colleagues, humans had nothing to do with it. At least this argument exonerates us from responsibility for the extinction of the largest Pleistocene mammal. It would be very convincing if not for evidence that ancestors of the Inuit were present on Wrangel Island 3,000 years ago. Is it possible that they arrived 1,000 years earlier and annihilated the very last of the woolly mammoths? Are we guilty after all? The only answer we can offer for now, as is so often the case in science, is that more research is needed.

PART THREE

The Storytellers

CHAPTER SEVEN

A Poisoned Gift

To know that we are mortal means that our lives are already lost,
no matter how many risks we succeed in avoiding. If animals were
aware of their mortality they would escape their zoological limbo;
they would stand erect.

Fernando Savater, Diccionario Filosófico

The Discovery

By the time Sima de los Huesos was populated, evolution had
spectacularly increased the size of the brain. As a result, there
were considerable advances in higher-level mental abilities and
an expansion of consciousness, the employment of which preceded an ever-increasing number of activities, for this faculty
was no longer limited to awareness of the present, but extended
to the future, to what was yet to come. Thus humans could anticipate events in the natural world and the behavior to be expected
of their fellows.

And then it happened. A sensational discovery, the first great
result of reflection and the prelude to all the rest; something that
each of us realizes at some point in life, since we are not born
knowing it. The hominids understood that all of them, without

exception, were going to die. This discovery was no more than the result of a very simple and purely logical analysis, but no other creature has made it. If all the others inevitably die, and I am not unlike them, then I too will die some day. In order to reach this conclusion, it is necessary to distinguish between *I* and *the others*. We attribute that capacity to *Homo ergaster* and perhaps to the Australopithecines. We do not know who first realized the inevitability of death, or where or when that realization occurred, but there is no doubt that the people who inhabited the Sierra de Atapuerca 350,000 to 400,000 years ago were aware of their own mortality. Ironically, more than 3.5 million years of evolution had produced exceptionally intelligent beings who came to realize that the course of a life is also a countdown to death. As Ecclesiastes 1:18 says, "For in much wisdom *is* much grief: and he that increaseth knowledge increaseth sorrow." Intelligence was a poisoned gift.

Many thinkers, Fernando Savater among them, believe that the understanding of our own powerlessness in the face of death's inevitability is what makes us truly human. If so, then the Atapuercans of 350,000 to 400,000 years ago must be accepted as full members of our aggrieved family, for there is no doubt that they were burdened with that terrible knowledge. But the same intelligence that brought us to the recognition of death also enabled us to understand for the first time that we are alive. Fernando Savater believes that the prehistoric humans who discovered death reacted by celebrating life. Through self-adornment and embellishment, they affirmed their existence, defying the final tragedy to come. They employed symbolism to express their immense joy at being alive (still). I am reminded of the words of a character in Amin Maalouf's novel *Leo Africanus*: "If death was not inevitable, man would have wasted his whole life attempting to avoid it. He would have risked nothing, attempted nothing, undertaken nothing, invented nothing, built nothing.

Life would have been a perpetual convalescence." Later I will discuss myth, symbolism, and art, but for now I want to look at the longevity of these prehistoric people who had discovered what awaited them at the end of life's journey.

The Life Span of Prehistoric Humans

People often comment to me that in prehistoric times life expectancy was very limited; people died quite young. When the cave paintings at Altamira were produced, they say, the few people who lived to be thirty were very old indeed. The first statements are partially true. The average age at death was much lower than it is in Spain today. But they are partially false. Not everybody died before the age of thirty. The second idea is completely inaccurate. Biologically speaking, the men and women at Altamira were no older or younger at the age of thirty than any of us is today. I hope that in my particular case I would still be able to keep up with a prehistoric group as they wandered in search of food, even though I celebrated my thirtieth birthday some time ago. Maybe I could even help them out a bit with their hunting and gathering. I don't see why I should be long dead if I were a prehistoric man. Perhaps I would be if I had had bad luck. All of us have been in a life-threatening situation at some point in life. But I might still be alive if I had enjoyed better than average fortune or if I could boast of outstanding survival skills.

Joking aside, in the following paragraphs, we will examine these rather complex topics, beginning with a look at modern hunter-gatherers. Many people assume that none of these people reach old age, either. But the data may surprise you.

When attempting a demographic study of a population that does not employ birth certificates or identity documents, a sometimes insuperable problem arises. It may surprise the reader to learn that people do not know even their approximate ages. The reason for this surprising (only to us) ignorance is that the age of an adult is of no importance. It is an irrelevant piece of data, of no interest to anyone. The stages of life are important. They are infancy, childhood, adolescence, and adulthood. Family relationships like mother, father, offspring, siblings, etc. are meaningful. But one's exact chronological age? No.

The only investigative technique we have to address this problem is to try to arrange the members of a family by birth order, beginning with the last born and working back to the first. In this way we can produce a table of relative ages. Still, it is not always easy to determine who is older than whom. One must always ask several people for verifying information. Even among siblings, the researcher may receive conflicting reports on birth sequence!

Once the members of a community are chronologically ordered, it is necessary to determine the exact age of at least some individuals, so as to estimate the ages of others who were born after one and before another. Sometimes this operation is possible, for example in the case of groups who have been visited by researchers at different times over the years so that individuals can be followed through life, beginning in their infancy. This process is made more complicated when people change their names at different stages of life, as they are wont to do. In the case of the Eastern Hadza, though, we have had considerable success. We have an acceptable 1985 census of 706 people that specifies the numerical ages of forty-eight individuals and provides relative ages for all the rest.

The Hadza are a contemporary hunter-gatherer people that speak a common language and live in a 1,000 square mile territory near Lake Eyasi in northern Tanzania. In recent years they have been studied by James O'Connell, Kristen Hawkes, and Nicholas Blurton Jones. They are one of the very few peoples with a nonproductive economy, that is who neither raise livestock nor cultivate the land, whose demography we have been able to study in detail. Another African people who practiced a similar economy until recently are the Dobe !Kung, who live in the northern Kalahari Desert in Botswana. Nancy Howell has done work on their demography. The Ache, a people of Paraguay, have only recently abandoned the hunting-gathering way of life. Their demography has been studied by Kim Hill and Magdalena Hurtado. Some other living people are primarily but not strictly hunter-gatherers. The Yanomamo of southern Venezuela and northern Brazil are an example, cultivating small areas of the forest, but still living largely from the hunt and by gathering wild fruit. They too shed light on the demography of populations without access to modern medicine.

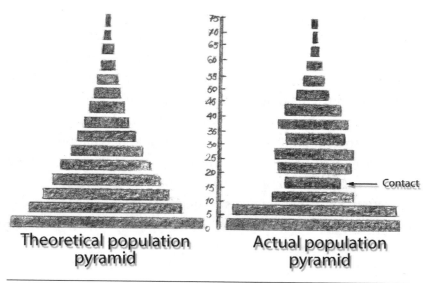

Figure 21: Theoretical (or ideal) and actual Ache population pyramids in 1987. The length of each bar corresponds to the percentage of the total population at the indicated age interval. Peaceful contact with neighboring peoples in the 1970s produced enormous mortality rates, mostly due to contagious diseases, which altered the population pyramid. Many children died either from disease itself or because their sick parents could not care for them. Data from Hill and Hurtado (1996).

Thanks to the data collected by researchers, demographic pyramids can be constructed for these peoples. A demographic pyramid illustrates the age structure of a population. It specifies the number of live individuals of each age interval, which may be of five years, for example 0-4, 5-9, 10-14, etc. The youngest children form the base and comprise the most numerous category. The bars above, representing successive age intervals, diminish in length as they ascend, since some people die at each age interval. As a result, the graph takes on the form of a pyramid.

In the Hadza demographic pyramid, the longest bar is of course the first, which represents all the children under five years of age. They make up about fifteen percent of the 706 individuals enumerated. Those under twenty are about fifty percent of the total population, those under forty are about seventy-five percent of the total, and those under sixty comprise a little over ninety percent. So almost ten percent of the Hadza are over sixty years of age, a very respectable age for people who hunt and gather for survival and who coexist with lions and hyenas. There are many Hadza children and adolescents, but no lack of adults and elderly people, either. Life is not as precarious for modern hunter-gatherers as some might suppose. The !Kung population of the Kalahari is somewhat older than that of the Ache. Individuals under twenty, for example, make up only forty percent of the population, not fifty percent as among the Hadza. This is because the !Kung population is stable, while the Hadza population is growing, so there are more young individuals.

Modern hunter-gatherers may not know their ages, but the deceased are even less likely to aid us in the collection of demographic data such as age at death. Fortunately, tombstones provide us with written evidence of the human life span in the recent past, which we can use to establish the mortality profile of historical populations. Just as a demographic pyramid illustrates the age structure of a live population, mortality statistics do the same for those who have passed over to the other side.

Several years ago, Antonio García y Bellido undertook to analyze the gravestones of Roman Hispania and produced a small set of statistics from that period in an entertaining and informative book called *Veinticinco Estampas de la España Antigua (Twenty-two Images of Ancient Spain)*.

About 5,000 funerary inscriptions were available to García y Bellido, representing the entire period of Imperial Roman Hispania, from the time of Augustus to the fall of the Empire, but primarily from the first three centuries A.D. In order to best represent natural mortality among civilians, he discounted the epitaphs of soldiers and people who died violently. In addition, he took into account only the deaths of people over ten years of age, so as not to consider the presumably high rates of infant and childhood mortality. Since so many gravestones were available, García y Bellido analyzed only two samples of 100 stones each, one from Lower Andalucía and the other from the Cantabrian coast. Within the sample from Andalucía, about one third of the individuals died between ten and thirty, another third between thirty and fifty, and the final third above fifty. This final third was evenly divided between those who died before the age of sixty and those who died at that age or above. One-half of the Cantabrians died between ten and thirty years of age and one-half above thirty. Although the latter sample exhibits a very high rate of juvenile mortality, much higher than in Andalucía, the longevity of those who survived past the age of thirty was also greater. Of the 100 Cantabrians in the sample, eighteen lived beyond seventy. In any case, these samples may be too limited to establish very rigorous comparisons between them.

These simple statistics do illustrate that despite what many say, not everyone died young in Roman Hispania. Life expectancy at birth, i.e. the average age of death for the entire population, is a very different question, accounting as it must for the egregious high infant and child mortality of the time, which dragged the average down dramatically. A cursory examination of dynastic genealogy suffices to get an appreciation for the latter factor. Of course the royals were also those who received the best available care as it was understood at any given time. As Gregorio Marañón used to say, the princes who eventually reached the throne were often the lucky survivors of a cataclysmic history of deceased siblings. Overall, life expectancy in the Roman Empire would have been about thirty years, varying somewhat by region. In some places it was even lower. Life expectancy is better, at about thirty-one to thirty-two years, among the Tanzanian Hadza today, which just goes to show that "civilization" did not always provide humans with a better life, or at least with lower mortality. The productive economy simply increased the number of human beings.

Once past the dangerous period of infancy and childhood, life expectancy in Roman Hispania was not so low. Antonio García Bellido provides the following data for Lower Andalucía and the Cantabrian coast: The life expectancy of ten year olds, i.e. the average period left to live for those who had reached the age of ten years, was thirty years in both regions. So those ten year olds could expect to live an average of thirty more years, to the age of forty. Not so bad. Then, for those who survived adolescence and reached adulthood, life expectancy was even rosier. The average age at death for European adults, that is, for people over twenty, seems to have remained constant at about fifty to fifty-five years until very recently. These relatively high figures should not be surprising. Upon reaching the age of twenty, a Paraguayan Ache woman can expect to live to an average sixty years of age, and a man to about fifty-four. The fact is that all of these demographic parameters changed little since the Neolithic, and perhaps since the Upper Paleolithic, until the middle of the nineteenth century.

At that time, the so-called demographic revolution took shape in the industrialized world, dramatically raising life expectancy at birth.

Even before the English country doctor Edward Jenner discovered vaccination against smallpox in the late eighteenth century, and at an ever-increasing pace since, the theoretical advances of modern medicine, the expanding availability of medical attention, and the wide application of measures for public hygiene brought down infant mortality rate to the point that the death of an infant or a young child is experienced as a terrible tragedy. But not many years ago, such deaths were accepted with weary resignation as predictable and daily occurrences. And of course the situation was much worse in medieval times, when horrific plagues, invisible enemies against which there was no known defense, swept periodically through Europe, decimating societies and killing millions.

Today's life expectancy at birth is over seventy years in the entire developed world and in some parts of the less-developed world. Still, it is under fifty in many African countries, and under forty in the most unfortunate among them. It is to be hoped that this situation will not last, because we now possess all the means necessary to correct it.

We have more fossils from the Neanderthals than from any other extinct human species, in principle a great advantage for studying their demography. But the paleodemography of pre-modern human species has always been one of the most challenging aspects of paleoanthropology. We are presented with two very difficult problems. One is that we do not have the fossilized remains of a single population, as in a cemetery. Instead we have partial skeletal remains of numerous individuals who belonged to distinct populations in very different times and places. In the case of the Neanderthals, there is a geographical range from the Iberian Peninsula and Wales to Uzbekistan and Iraq, and an age range of many thousands of years. Of course to some extent we can say that all of these individuals comprise a single sample that may be taken to represent average Neanderthal mortality. But if we consider that no demographer would dare to analyze a sample that included the remains of the modern Spanish population together with those of the Palestinian Jews of the Roman Empire, we get an idea of what it would mean to lump together data separated by thousands of years and representing people who lived in very different environments and circumstances. There is no immediate solution to this problem. Paleontologists are always constrained to work with the fossils available to them, and at the same time to continue adding to that sample in the hope of increasing its representativity. Every new discovery is a step in that direction.

The second conundrum is diagnosing the age at death of human fossil remains. If it is almost impossible, as I have mentioned, to determine with the necessary precision the age of living Hadza, !Kung, or Ache, how can we determine the age of Neanderthals who died thousands of years ago? There are two solutions to this question, neither of them fully satisfactory. One is to improve our technical capacity to determine the age at death of skeletons, and the other is to group the fossils in several broad age categories, compensating for the imprecision of the categories with the increased breadth of the sample. Sometimes the age at death is simply indeterminable. Then we must be content to know the decade or an even less precise period of life in which an individual died. The age at death can be determined quite accurately for individuals of our species who had not completed their development, i.e. if all their permanent teeth had not yet come in, or if their bones had not fully knit. We have detailed tables of tooth eruption, which is the formation and growth of the permanent teeth, and of the fusion of the epiphyses, which is the definitive knitting of the bone extremities.

When the subject is mature, no more teeth will be coming in and the bones are no longer growing, so we need to employ other means to determine the age at death. Several have been tried. One means that was formerly very popular was to examine the extent of fusion at the sutures that connect the cranial bones. In individuals who have not completed their development, these sutures are open and the bones of the cranium can still separate. These only knit together completely with time and maturation. This method was eventually discounted, however, because the sutures do not close at a universal or regular rate. Other scientists have tried, without much success, to determine age through an examination of microscopic bone structures, or bone histology.

Today there are several commonly used methods. One is based on certain modifications that take place at the hip and that affect the surfaces of the two coxal bones where they articulate with the sacrum, what are called their auricular surfaces. Similar processes occur at the pubic symphysis, where the two pubic bones closely approximate each other at the most anterior point of the pelvis. Additional changes associated with aging take place in adults in the trabecula, the internal structure of the upper femur and humerus. These changes can be detected by means of radiography.

Another method that is often employed when there is no other option is an analysis of tooth wear, specifically of the crowns of the molars. Naturally, the teeth of older adults are more worn than those of younger ones. But tooth wear depends on diet, especially on the presence or absence of abrasive particles in the food, so the analytical parameters of tooth wear must be separately calibrated for each specific population. This is done by calculating tooth wear among nonadults as it corresponds to the eruption of their permanent teeth and then applying the same rate of wear to adults to estimate their ages of death. All of these methods generally work better with younger adults than with older ones. As the age of the subject increases, the validity of our estimates decreases.

To apply these criteria for estimating the age at death to the fossils of other human species, we must assume that the pace and stages of their physical development and aging corresponded to ours. In the case of the Neanderthals, most researchers believe that assumption is justified, given the close relationship between brain size and development. Since the average Neanderthal brain size was equal to our own, we suppose that their development was not much faster than ours. It has also been demonstrated that the sequential stages of their development essentially corresponded to those of modern humans, so it does not seem unreasonable to apply our standards of aging to them as well.

Erik Trinkaus, today's leading Neanderthal specialist, has entered the ages at death of all known Neanderthal individuals onto a paleodemographic table. Two hundred six individuals are represented, a truly impressive number that attests to the extensive information now available to us about this extinct human race. All of them lived in the Upper Pleistocene, less than 127,000 years ago. But we do not have 206 complete skeletons. We have parts, sometimes very few parts, of 206 skeletons. Trinkaus divided the individuals into six broad categories: 1) Neonates: individuals under one year of age. 2) Children: over one and under five years of age. 3) Juveniles: over five and under ten years of age. 4) Adolescents: over ten and under twenty years of age. 5) Adults: over twenty and under forty years of age. 6) Old adults: forty or more years of age.

There are very few neonates in the combined Neanderthal sample, which does not surprise paleodemographers. After all, small children are frequently underrepresented in necropoli throughout the ages. This may be in part due to the fragility and ease of decomposition of these particular bones, but in many cultures, very young children have simply not been considered "people," and their bodies were not deposited in the cemeteries with their elders.

By analogy with contemporary populations sadly lacking access to modern medicine and with those who lived prior to the demographic revolution, we know that the mortality of children under five would have represented some forty percent or more of the total. In other words, nearly half the population would have died before reaching the age of five. Mortality would then have dropped among juveniles and adolescents, but risen again among adults. This U-shaped pattern is common to all mammal populations. The probability of death is greater among very young and very old individuals than among those who have survived the appallingly dangerous period of neonatal mortality and the crisis of weaning, but who still enjoy the care of their parents and have not yet had to face the dangers of adulthood. When I say "the probability of death," I am using a demographic term that means the probability of death within one year or some other defined period. Statistically, I fear, the overall probability of death is one hundred percent for all of us.

However, there is one surprising aspect of the Neanderthal data. There are many fewer "old adults" (over forty years old) than would be expected. In comparable modern populations, about one half of the individuals who reach adulthood also reach old adulthood, while among Neanderthals, only twenty or perhaps thirty percent at most of all adults survived past the age of forty. If we estimate child mortality (mortality among children under five years old) at between thirty-five and forty-five percent, we find that the number of Neanderthals who died at ages of forty and above would have represented only about six percent of the total population. This anomaly requires some explanation if we are to believe that potential Neanderthal longevity was equal to ours. Even among chimpanzees, whose potential longevity is about half ours, thirty-five percent of the population dies after the age of twenty-seven, which Trinkaus considers comparable to the age of "old adult" humans.

One possible solution would be to accept the data and conclude that the life span of Neanderthals was much shorter than that of modern humans, including those modern humans with a lifestyle that would appear to be comparable to the Neanderthals'. This could be because the Neanderthals were faced with such dangerous circumstances that few of them survived past forty. Although it is reasonable to accept the idea that Neanderthal life expectancy at birth was well under thirty years, the almost complete absence of "old adults" requires a scenario of demographic stress so acute that it would become insuperable under any unfavorable circumstance, for example an ecological crisis resulting from drought, unusually long or harsh winters, an epidemic among game animals, several years of low wild fruit production, etc.

With such high mortality and such low life expectancy at birth, a very high rate of fertility would have been required to make the Neanderthal populations demographically viable. Fertility among modern hunter-gatherers is quite variable. !Kung women have an average of 4.7 children, the Hadza have 6.15, and the Ache have about eight. But if Neanderthal mortality was much higher than that of modern groups like the Ache, among whom the average age of death for men is about fifty-four years

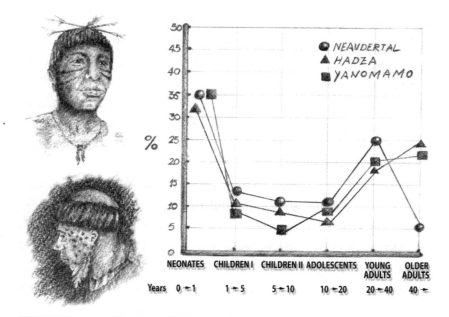

Figure 22: *Mortality Rates among the Hadza, the Yanomamo, and the Neanderthals. The vertical axis expresses the percentage of the population that died at the stages of life indicated on the horizontal axis. Data taken from Trinkaus (1995).*

and for women about sixty, then their fertility would have needed to be even greater. One way to increase fertility is to increase the frequency of pregnancies, but it is difficult to imagine that the Neanderthals could have outdone the modern populations in this department. That would have implied earlier weaning, which of course would have meant the weaning of children even less well prepared to survive the associated crisis of mortality. Among the !Kung the average interval between births is about four years and among the Ache it is approximately three years.

The reproductive period also ends earlier among !Kung than among Ache women, which may be due to the prevalence of sexually transmitted diseases among the former. Neanderthal women could have increased their fertility with a longer reproductive period, not by extending it, but by beginning it at a very early age. But their age at first conception and birth would have had to be much, much lower than among our species in order to compensate for their extraordinary mortality.

Many people believe that the age of menarche, or a woman's first menstruation, has increased in the developed world since the industrial revolution. In fact it has fallen. This is because the onset of the female reproductive period is strongly conditioned by the quantity and quality of food available during a young woman's preceding developmental periods. A theoretical model can be devised to relate the age of menarche and of first childbirth to the body weight of girls, which itself reflects nutritional adequacy. Among hunter-gatherers, who are intensely active and consequently burn a lot of calories, yet whose caloric intake is limited, the onset of fertility is generally later than it is in the developed world, where the availability of calories may be unlimited. Among Ache girls, the most frequent age, or what statisticians call the "mode," for menarche is thirteen years. Among the !Kung, who are significantly less well nourished, the age of first menstruation is typically seventeen. Our statistical model would predict that the first childbirth of the young women of these two groups would be at eighteen and nineteen years respectively, and indeed the real world figures are not far off at seventeen and nineteen years. Among well-nourished white girls in the United States, on the other hand, the model predicts that the first menstruation will be at the age of twelve and the first childbirth at sixteen. The first figure corresponds to reality and the second clearly does not, but that is for cultural rather than biological reasons, i.e. sexual abstinence or the use of contraceptives. In any case, even if we were to posit the early age of sixteen years for the first childbirth of Neanderthal women, it would not compensate for their puzzling mortality. It is also absurd to imagine Neanderthal girls living the sedentary life or receiving the nourishment that white American girls do.

In addition, given that Neanderthals developed slowly just as we do, the early death of parents would typically have left several orphans still very much in need of care. No, there must be another explanation for the seeming absence of old Neanderthals.

Perhaps the criteria being used to calculate the ages of death are inaccurate, and we have systematically classified older adults as younger ones. It is possible, for example, that beyond a certain level of tooth wear or modification of the symphysis pubis, the signs of aging may stabilize and cease to reflect age proportionately. In that case we may be mistakenly categorizing some individuals over forty years old as young adults. The two possibilities contemplated up to this point, high mortality and erroneous age calculation, are reasonable and may even have combined to give us these surprising results, but Trinkaus advances a third explanation that I find even more intriguing.

All but four of the Neanderthal fossils in the sample have been found in caves. If we imagine these caves to have been the habitual residences of the Neanderthals, their dwelling places, then we should have found the old adults missing from the paleodemographic chronologies right inside them. In fact, we do find many signs of the use of caves as dwelling places by Neanderthals. They left enormous numbers of lithic tools, stone chips from their production, and the bones of food animals. But in the very protracted period of time during which these objects accumulated, the caves took on the characteristics of palimpsests, aecheological documents upon which numerous episodes of cave occupation were superimposed, one atop the other. Each such occupation, though, may have been of very limited duration, perhaps minutes, hours, a few days, or a couple of weeks. They may also have been separated by long periods of time, perhaps years, decades, centuries, or even millennia. We cannot discern the texture of time and events within a given archeological stratum. The record is written on a single plane, so events perceived today as contemporaneous were in fact widely separated in time. We do not have access to any temporal-sequential perspectives within individual strata.

What follows is definitely my favorite explanation. The Neanderthals' principal dwelling place was under the stars. The caves were just another element of the natural environment, a resource to be used occasionally and for short periods of time, principally for refuge. Nevertheless, the caves' geological characteristics made them the principal repository of Neanderthal fossils, practically the only one in fact, thus distorting our perception of their lives. Neanderthal life was very probably characterized by significant mobility. With this in mind, it is fair to say that the deaths of old adults would have been much more likely while ranging through their wider habitat than when sheltered and at relative rest in caves.

Many Neanderthal remains show signs of having suffered from disease or trauma in the course of their perilous lives. The famous Old Man of La Chapelle-aux-Saints was a classical Neanderthal if ever there was one. When he died, he suffered from generalized arthritis, possibly the result of trauma, and had lost nearly all his teeth. By the way, he was not really so old at the time. Trinkaus estimates his age at death at about thirty. Other Neanderthals also suffered degenerative joint diseases and numerous bone fractures. Individual One at the Iraqi site of Shanidar probably lacked the use of his left eye as the result of a violent blow. His right arm had been amputated above the elbow, and he had received additional strong blows to the right hip, ankle, and foot. The fact that he survived all these injuries indicates that he was cared for by his group.

Erik Trinkaus and Tomy Berger have found a lot of similarity in the distribution of injuries on the bodies of Neanderthals and of rodeo professionals. Today's daring rodeo cowboys are violently thrown to the ground by horses and steers. They are most often injured on the head, trunk, and arms. Neanderthal hunters were obliged by circumstances to approach large, powerful animals very closely, presumably with the same results.

Yet despite the large number of injuries sustained by the Neanderthals in the course of their hazardous lives, not one of the individuals represented in the fossil record had lost all mobility in his or her legs. Though some were beyond hunting, all retained the ability to walk. Perhaps they were fed by the group, but in every case they were ambulatory. If the caves were but stops along the way, shelters within a broad area where a given group perambulated, then perhaps most older individuals simply fell by the wayside between one cave and another or between two visits to a single cave. Some combination of this hypothesis with the two explanations suggested above, i.e. age-determination methods that "rejuvenate" oldsters and a very limited life expectancy, could be the definitive answer to the scarcity, if not absence, of old Neanderthals at cave sites.

What Happened at Sima de los Huesos?

Sima de los Huesos, literally the "Bone Pit," is a wonderful exception to the rule that very few people are represented at any given find of Neanderthals or their ancestors. The Atapuerca Project took shape thanks to the discovery of a handful of human remains at the Sima in 1976. Some years later, Sima de los Huesos has yielded more than four thousand human fossils, even though only a very limited area has been excavated and to only a shallow depth. No other archeological site has produced such a wealth of fossils of the genus *Homo* that are older than the modern human burial sites of the late Upper Paleolithic. Not only has Sima de los Huesos provided an enormous number of human fossils, but these represent every part of the skeleton, including the smallest bones of all, the bones of the middle ear, the malleus, the incus, and the stapes, popularly known as the hammer, the anvil, and the stirrup. In other sites where we have been fortunate enough to find human fossils, only fragments of crania and mandibles have turned up. From the time of the Australopithecines and the paranthropoids to the time of the Neanderthals, few skeletal remains were found. They are appearing now in Sima de los Huesos because many bodies accumulated in this cave, and their intact skeletons were preserved surprisingly well, despite the intervening 350,000 to 400,000 years. The retrieval of all the bones belonging to these skeletons is only a matter of time and patience.

To date, the deposit of human remains at Sima de los Huesos has been probed in only a few places. The objective of these probes has been to get an idea of what the site contains and to understand it better in order to make the best possible plans for its excavation. Since the bodies are stacked one on top of the other, the probes disclose the separate bones of different individuals instead of complete skeletons. For example, we may recover the arm of one individual on top of the hip of another on top of the cranium of a third and so on. Little by little, over the course of many years and depending on how the excavation is accomplished, we will get a better idea of how the remains are arranged. For now, the best way to identify individuals from among the many sets of remains is by their dentition. It is easier to associate teeth with other teeth from the same set than to try to associate individual bones with other bones from the same skeleton. In addition, every adult has thirty-two teeth, a good number to work with, and they are perfectly preserved at this site.

José María Bermúdez de Castro is working closely with the teeth at the National Museum of Natural Sciences in Madrid. He is also one of the three directors of the Atapuerca Project, along with me and with Eudald Carbonell at the Universitat Rovira i Virgili in Tarragona. At this time, José María Bermúdez de Castro has identified twenty-eight individuals on the basis of their complete or partial dentition. This is actually a minimum, because there are many individual teeth that may or may not have belonged to one of the same thirty-two people. He will probably identify more people on the basis of these teeth and more still as we continue to work at the excavation. Teeth often tell us the age at death of their former owners, very precisely in the case of those at intermediate stages of development and less dependably so in the case of adults, since the analysis of crown wear is less exact, especially for individuals of advanced age.

The analysis of teeth recovered so far at Sima de los Huesos tells us that there were few "old adults" there. Only three individuals were only than thirty. This is not surprising, since it corresponds to what we have found among Neanderthals and other fossils from the period of Sima de los Huesos. Individuals exhibiting advanced tooth wear are always scarce. The majority of the sample did not have the occasion to chew very much before they expired. Of course one of the "old adults" is our Cranium 5, the most complete of the collection and of the entire human fossil record. But "dental age" must be analyzed in the light of other indicators of age at death. There are two pubic bones at Sima de los Huesos that belonged to men over forty. One of them indicates that the person in question was considerably older, definitely over fifty, in fact. The Sima de los Huesos collection also includes a pubic bone that belonged to a woman of about thirty. These three pubic bones and their corresponding pelvises may have belonged to the same three "old adults" identified on the basis of their dentition.

The pubic evidence indicates that the potential longevity of this population was equal to ours, so the scarcity of individuals over the age of thirty must have another explanation. Perhaps the circumstance suggested for the Neanderthals applies here as well, that the people of Atapuerca moved around the area a great deal and only took shelter in mountain caves from time to time. By the time the oldest individuals found themselves unable to cope with the demands of life, they were not strong enough to reach the caves, and literally fell by the wayside. When a member of the group died in the caves at Sierra de Atapuerca or nearby, the body was carried to this hidden niche and deposited there. That is why we came to call this place Sima de los Huesos, which in English means bone pit or depository, in other words, an ossuary. This burial tradition would probably have been maintained by a particular human group for several generations, perhaps until the custom was forgotten or the group that had practiced it disappeared. But by that time, the Sima de los Huesos was home to an impressive collection of human skeletons.

Of course there is also the very interesting possibility that all the Sima de los Huesos humans died together or within a short period of time. How can we determine whether these twenty-eight or more sets of remains accumulated in a short time or over an extended period? From the geological point of view, hundreds or even thousands of years are no more than the blink of an eye, so a stratigraphic analysis would not answer the question. All the fossil remains are found in the same sedimentary layer, encased in the same deposit, not in successive ones.

What we can do is to disregard all subjects twenty-five years old or more, of whom there are always very few in the fossil record, and children under five, who are also underrepresented at the Sima. Only one child has been tentatively identified in the group, the poor thing having died between the ages of four and six. My good friend Jean-Pierre Bocquet-Appel of the Museum of Man in Paris, one of today's most brilliant paleodemographers, proposed the following test for individuals between five and twenty-four:

In known populations without access to modern medicine or living prior to the demographic revolution, there are about as many living people between five and fifteen years of age as there are between fifteen and twenty-four. Actually there are a few more, about 115 percent as many: (5-14 / 15-24) x 100 = 115%. Nevertheless, in the cemeteries of these same communities, the former group is represented more than two times as frequently as the latter: (5-14 / 15-24) x 100 = 225%. The reason for this discrepancy is that the demographic pyramid of a community, which represents the age structure of its live population, provides different information than its mortality statistics, which describe the distribution of mortality by age. Jean-Pierre Bocquet-Appel proposed to determine which of the above mentioned two percentages better described the deceased population at Sima de los Huesos. It turned out to be fifty-three percent, a proportion much closer to that of the demographic pyramids than to that of the mortality statistics.

What does this result tell us? Well, the age distribution at Sima de los Huesos is suggestive of a catastrophic event that killed many members of a group in a short time, rather than the slow generational attrition that a population experiences under most circumstances. Indeed, the proportion of fifty-three percent found at Sima de los Huesos is too low even to correspond to a demographic pyramid. There are very few children or young adolescents. In fact the strongest, most mobile, and most active members of the population predominate. So what happened at Sima de los Huesos? What was the nature of the catastrophe?

The possibility of an epidemic is one thing that immediately comes to mind. But the existence of large-scale epidemics in prehistory has always been considered improbable due to the small human population and the low population density. Most pathogenic agents are very short lived and cannot spread if host communities are small and have only infrequent contact with one another.

Hadza population density is about 0.8 persons per square mile. This density is considered quite high, but it is made possible by the high level of animal and vegetable biomass in their environment. Expressed in other terms, we can say that a band of thirty-one Hadza move around within an area of about forty square miles, the equivalent of a square measuring about 6.3 miles on each side. A similar or even greater population density has been attributed to the late Upper Paleolithic populations of the Mediterranean forests on the Levantine coast during the mildest climatic periods. Ecosystems producing less biomass would support sparser populations of hunter-gatherers. Among the Dobe !Kung and the San, who live in the Kalahari Desert, population density may be as much as ten times less than among the Hadza, just 0.08 persons per square mile. That is to say, the same band of thirty-one individuals would hunt and gather within an area equivalent to a rectangle of fifteen miles by nearly twenty-seven miles. Multiplying these two population density figures by the area of the Iberian Peninsula, about 230,000 square miles, we can get an approximate idea of its prehistoric population, between 18,000 and 180,000 human beings.

The technology employed until almost the end of the Pleistocene was far less sophisticated than that used by today's hunter-gatherers, so the figure of 18,000 is surely more realistic, especially during colder periods when environmental conditions were very difficult. But a number of 180,000 or even more is conceivable for the Mesolithic, the period of the Holocene just before the invention of agriculture and animal husbandry, which would increase population density even more. By the time of Columbus' voyages to the Americas, the peninsula would have had about seven million inhabitants.

The estimates of human population density in the Paleolithic could be compared to the density of other present-day mammal species on the peninsula, were it not that environmental degradation has left few large areas within which herbivores and carnivores survive in a more or less natural or balanced state. One of these few refuges is at the Sierra de la Culebra in Zamora, a 165,000 acre game reserve. The wolf population is protected there, and the reserve is home to what is possibly the densest wolf population in all of Europe. According to the research of José Luis Vicente, Mariano Rodríguez, and Jesús Palacios, wolf population density on the reserve ranges from 0.13 to 0.26 individuals per square mile, or from five to ten individuals in a forty square mile area. So the population density of these social hunters is between that of the !Kung and that of the Hadza, and our speculation about the number of prehistoric Iberian hunters does not seem implausible.

In the areas of deer population, their density reaches one per square mile, or forty animals in that forty mile square. The average deer weighs 400 pounds. The number of animals hunted by humans each year is very limited and does not affect the size of the population.

If the human population was so small in the Pleistocene, one may legitimately wonder how at least thirty-two individuals could die together. Taking into account that some group members must have survived in order to deposit the bodies of the deceased at Sima de los Huesos, and given that neither very old nor very young individuals are among the remains, it seems that to support the catastrophic hypothesis we would need to posit an impossibly large group. After all, many experts believe that humans lived in very small groups at the time and that these groups remained isolated from each other.

Jean-Pierre Bocquet-Appel has analyzed the size of prehistoric human groups from an unusual point of view. The sex of an individual is a kind of variable that in statistics is called binomial, meaning that there are only two alternatives, in this case either male or female. The probability of arriving in the world as a person of one sex or the other is approximately equal, since about 105 boys are born for every 100 girls. This is not to say that every couple with four children will have two girls and two boys. That may be the average, but we all know that there are many families with only sons or only daughters. If we consider a community of, say, twenty couples instead of thinking about individual families, we find that the probability of an all-boy or an all-girl generation is minimal. Nevertheless, we can also be quite sure that sooner or later there will be a generation with very few boys or very few girls. The smaller the population, the greater the generational variation in the proportion of males compared to females.

It is relatively easy to construct theoretical models to study this problem. Jean-Pierre Bocquet-Appel has done so, and he has concluded that in the long run, small groups would not be able to survive without exchanging either men or women with other groups in order to maintain sexually balanced populations. In fact, if a group had twenty members between the ages of fifteen and forty years, the migration rate would have to average eleven percent, meaning that about two members of the majority sex would have to leave the group and about two members of the opposite sex would have to enter it. If the groups had fifty members, the rate of migration would need to be seven percent, meaning that three or four people would need to be replaced. Finally, if the group had 350 to 400 members, three percent or about ten to twelve people would have to migrate out, etc. Groups participating in such exchanges are said to be interrelated by means of exogamy, forming a much wider group for the purpose of reproduction.

According to Jean-Pierre Bocquet-Appel, Acheulean and Mousterian industry were practiced over wide ranging areas for precisely this reason. The human population never reached a density high enough to permit the existence of demographically self-sufficient or culturally isolated local groups. Instead they formed a very extensive network of small groups that were genetically and culturally interconnected over long distances. At times they would have tended to come together in larger units, and at other times they would disperse and live in small encampments.

Robin Dunbar uses another method to determine the size of human groups. He has studied the relationship between the size of the neocortex in primate species and the size and complexity of their social groups. Based upon the size of the human neocortex, we should ideally have direct relationships and establish personal ties with a group of 150 individuals, although we would not necessarily have to be together all the time. In keeping with Jean-Pierre Bocquet-Appel's demographic model, people who did not mate within this limited number of friends and relatives, what we can call a "clan," would be able to seek partners from among neighboring "clans" of the same "tribe." Seen in this light, the twenty-eight individuals at Sima de los Huesos are not so numerous as to invalidate the possibility of a catastrophic demise. All of them could have belonged to the same one or two "clans."

While widespread epidemics like the plagues of medieval Europe would not have been possible in the period of Sima de los Huesos, a contagious disease could have affected one or more small, human groups. But the age profile of our twenty-eight individuals at Sima de los Huesos discounts that hypothesis. In two modern and therefore well-documented epidemics of cholera and smallpox, most of the dead were under ten years old, forty-five percent in the first case and ninety percent in the second. Epidemic disease generally kills more young children than adolescents and young adults, but it is precisely the latter two groups that predominate at Sima de los Huesos.

Jean-Pierre Bocquet-Appel and I are thinking of another kind of catastrophe, an ecological crisis. Life in the natural world does not lack alarming circumstances. In fact it is anything but stable. Animal and plant communities are subject to cyclical changes in the physical environment. These are generally minor, but from time to time there are long periods of unusual heat, drought, or several years running of particularly long and cold winters. In exceptional circumstances, these crises may be more severe or protracted. Not long ago we had a series of dry years in Spain that raised serious concern. Animal populations are very sensitive to these environmental fluctuations. Their numbers dwindle in challenging times only to surge in periods of abundance. Corresponding increases and decreases in predator and prey populations have been recognized since the first days of ecology as a scientific discipline. In a severe crisis, plants, herbivores, and carnivores all die in the affected region, as do humans. Ethnographic studies of modern hunter-gatherer populations have demonstrated that these calamities bring them great suffering. A nonproductive economy is dependent upon the availability of needed resources in the environment, so its practitioners must adapt to and endure whatever conditions may present themselves.

But human groups do not wait passively for such crises to pass. They take off in search of better circumstances. The weakest members fall by the wayside: children, the aged, the sick, the disabled. Ultimately there is an age-based selection process, since robustness corresponds to adolescence and young adulthood. This may have occurred 350,000 to 400,000 years ago on the Meseta and possibly also in the Ebro depression and in other nearby regions of the peninsular interior. When human groups set out for more favorable areas, one of the places they took refuge was in the Sierra de Atapuerca, an environment privileged by the particular ecological and geographical characteristics that I described earlier. The fossil beds found in several of its caves testify to a continuous human presence for at least the last million years. The strongest individuals reached this mountain refuge after a difficult journey, leaving many fallen comrades along the way. Once they arrived, their suffering continued for a time, or perhaps many individuals arrived in such a debilitated state that they just did not last much longer. In either case, many more died. The lucky survivors sought an isolated spot in which to deposit the cadavers of their deceased and thus protect them from the depredations of carrion eaters. They found a spacious cave with only one, very narrow mouth for ingress, which admitted virtually no light. Because of its difficult access and its lack of light, the cave had never been occupied by humans, although bears used it year after year for hibernation. In one corner not far from the entrance, there was a mysterious vertical shaft almost forty-six feet deep, although its bottom could not be discerned from above. It was here that they let drop the bodies of their departed in what was, as far as available evidence tells us, the first human funerary activity. The ecological crisis passed. Animal and human populations recovered. Life went on as it had for ages in the lands of the peninsular interior. But in Burgos, there was a cave that contained the remains of at least twenty-eight humans who lived 350,000 to 400,000 years ago. At some point the entrance to the cave was blocked by natural causes. Bears no longer entered it for their winter hibernation. In fact, nobody visited the Sima de los Huesos again until it was rediscovered by humans in the twentieth century.

Finally, the Sima de los Huesos recently provided us with a new surprise: a stone axe, the only tool found among thousands of human fossils of the skeletons of tens of individuals. It is a beautiful tool, carved from red quartz; its colour is rare in the river stones used by the prehistoric men of Atapuerca. We have called this stone axe Excalibur because, like King Arthur's magic sword, it seems to us a highly symbolic object. Perhaps it will be the definitive proof that something akin to a funerary rite was performed in that extraordinary place.

CHAPTER EIGHT

Children of the Fire

Or is the philanthropist or the saint to give up his endeavors to lead a noble life, because the simplest study of man's nature reveals, at its foundations, all the selfish passions and fierce appetites of the nearest quadruped? Is mother-love vile because a hen shows it, or fidelity base because dogs possess it?
Thomas H. Huxley, Evidence of Man's Place in Nature.

The Mind of a Hamster

My children have a hamster in a cage. It was born in another cage, like its parents before it. It belongs to a line of domestic animals many generations long and would be unable to survive in the wild. It isn't even the same color as its wild cousins. It is white, an albino we can say, whose color would quickly attract its natural predators. When my children give it a few grains of food, the animal stuffs them into its mouth as quickly as possible, but it does not swallow them. Instead it stores them in its buccal pouches, which are folds of skin that form natural pockets on the inside of its cheeks. The first time they fed the hamster, which was an adult when they got it, the children called me in alarm because its face was so swollen with stored grain that they

235

thought it was terribly sick. After filling its pouches, the hamster invariably goes to the other side of its little cage, where it has built itself a tangled nest of straw. There it expels the seeds and sets to eating them if it happens to be hungry.

With these curious habits, my domestic, but not domesticated, hamster reproduces the behavior of its ancestors in central and eastern Europe. On the steppes, a hamster that feeds nocturnally upon the seeds of gramineae runs a strong risk of ending up in the stomach of an owl. In order to minimize its exposure to this danger, the feral hamster fills its "grocery bags" just as quickly as it can by storing grain in its pouches and scurrying directly back to its underground burrow, where it can be safe from preda-tors and where it maintains a large store of food reserves. Our hamster, although it cannot dig through the metal floor of its cage, does the best it can to simulate a burrow with the materials at hand. It really should not fear for predators in our house, and it should realize that it will never lack for food. Nevertheless, it will never eat its grain immediately. It must first carry it to an imaginary refuge from an imaginary threat.

It is clear from observations like this one of the hamster's eat-ing habits that mammals have unalterable genetically pro-grammed behaviors that they are compelled to obey. A given stim-ulus automatically produces a predictable response behavior. As far as these rigidly programmed behaviors are concerned, animals are slaves to their genes. In another sense though, we can say that animals like our pet hamster are born with knowledge. Of course the behavior of this hamster that we observe in a cage makes no sense in its present condition as a domestic animal, but it would be perfectly "logical" in a natural environment, where an animal must do exactly what is called for to improve its chances for sur-vival and reproduction. For precisely this reason, natural selection has favored this particular behavior over the many other possible behaviors, just as it has favored certain body structures over others. But the hamster's white color is nonadaptive. It is not motivated by any natural advantage, and it would not be useful in a natural environment. In fact, it would be highly prejudicial to the hamster's survival. Though it originated as albinism, a natural and spontaneous mutation, it has been selected as preferable by humans. But the point here is to make it clear that animals have innate knowledge, unconscious of course. They do not come into the world absolutely ignorant of what awaits them. In a sense, their genes bequeath them the "wisdom" that every little hamster clearly possesses. To wit: It is useful to stuff its cheek pouches with all the food it can hold, to carry that food to its nesting place, and to consume or store it there and only there.

Earlier in this book, in a discussion of how Neanderthals may have reacted to the Cro-Magnons' physical appearance, I mentioned a triggering stimuli that acts upon us humans as well. I said that round and disproportionately large heads, high rounded foreheads, little bitty noses, and chubby cheeks stimulate our protective impulses, whether these features are observed on a real child or caricatured on Bambi. Since humans are primates and therefore visually oriented animals, it is easiest to identify triggering stimuli of the visual variety, although they exist for other senses as well. It is clear that olfactory stimuli are very important to other animals. When I explain in class that certain stimuli produce innate responses among us humans, I draw two circles on the board, and I put two dots inside each one. Below one set of dots I draw an arc open towards the top, and below the other set of dots I draw an arc open towards the bottom. Based on these few strokes of chalk, every student can identify a happy face and a sad face. Technically, I teach human paleontology, but this should help you to understand that we paleoanthropologists are interested in more than just fossils.

Of course, an animal's behavior is based on more than just the creature's genetic inheritance. Throughout their lives, animals also learn from the information available in the environment. This is especially true in the case of those animals with the most highly developed nervous systems, the mammals. To be explicit, we identify two categories of knowledge. Phylogenetic knowledge is accumulated throughout the course of evolution and is genetically recorded, while ontogenetic knowledge is accumulated by an animal during its lifetime. The human capacity for language allows us to pass along ontogenetic knowledge as cultural information. Life's positive and negative experiences are forever recorded and associated with certain places, animate and inanimate objects, and other circumstances. Who, for example, does not remember the aromas of his or her childhood?

The famous experiments conducted at the beginning of the twentieth century by the Russian scientist Ivan Pavlov were based on precisely this ontogenetic learning. Pavlov provided food to a dog at the same time that he rang a bell until the sound of the bell by itself became sufficient to stimulate the animal's production of saliva. Pavlov had produced a conditioned reflex based on the association of a particular stimulus with a positive experience. By the same token, a dog can be negatively conditioned to fear the sight of a whip. We have all seen a domestic dog almost miraculously divine when it will be taken out for a walk based on certain preparatory behaviors that its person regularly performs. Of course the most obvious of these is the taking up of the leash. "What a smart animal!" people say. But in Pavlov's experiment, the dog established an association between its feeding and a completely neutral and arbitrary signal, the sound of a bell, whose only relationship to food was temporal, i.e. simultaneity. The dog could just have easily have been taught to associate its feeding with the appearance of a large poster upon which the word "food" had been written. This would not have meant that the dog had learned to read. We will return to this shortly.

Both innate and conditioned reflexes turn animals into mere automatons, reacting to stimuli based on their genetic information or either positive or negative conditioning. It is often said that humans are the only animals that kill for pleasure, while even the most ferocious carnivores are more respectful of life and kill only to eat. Actually this is not quite true. Carnivores also hunt without necessarily being hungry, as anyone with a house-cat can tell you. Housecats will stalk anything that moves, stealthily approaching and finally pouncing upon it. If no animate prey is available, they will play at hunting with a toy. One could say that cats kill out of hunger but hunt "for pleasure." This last phrase is in quotations because it is a human concept. But in order to live contentedly, cats must engage in hunting behavior a certain number of times per day, whether they are hungry or not.

Based on these simple observations, we can say that animal behavior is subject to the creatures' reactions to external and internal stimuli, endogenous mechanisms that ethologists, the scientists who study animal behavior, call impulsions or drives. Thus animal behavior is not always reflexive. It may also be spontaneous, internally motivated that is. Intrinsic impulsions in animals produce physiological states that we can call (somewhat infelicitously) "moods." These "moods" produce internal tensions that drive animals to actively seek out the stimuli that will then trigger their own tension-relieving behavior. The longer the time elapsed since the behavior in question was last produced, the greater the tension and the less intense the stimulus needed to trigger it. In the ultimate instance, the behavior may be produced in a vacuum, with no real external stimulus. Sometimes the physiological basis for these drives is clear. For example, certain hormones are directly related to the sexual impulse. But in other cases the physiological factors are murkier.

Konrad Lorenz referred to these phenomena in his book *On Aggression,* which triggered adverse reactions from many academic psychologists and sociologists and from professors of education. But it primarily upset those who had not taken the trouble to read the book. They really should not have gotten so exercised. The existence of drives, including aggressive ones, is the norm among animals, but so is the existence of aggression-inhibiting stimuli. In addition, aggression can be rechanneled. In any case, I have to admit my admiration for Konrad Lorenz who, in the modern age of highly complex and expensive laboratory equipment, won the Nobel Prize by observing geese and jackdaws in his back yard.

Getting back to the point, animal behavior can be explained by the interplay between impulses and reflex behaviors, either innate or acquired. There does not seem to be much of a place for consciousness, and in my opinion animals do not experience anything similar to human consciousness. Of course, since there is no way to communicate with animals and ask them about what goes on inside their heads, we cannot directly know whether or not they have any degree of consciousness. That is why I approach the problem like this: Can we explain animal behavior without recourse to the idea of consciousness? If, as I believe, the answer is yes, then it is best not to attribute anything to them that does not fill an explanatory lacuna.

Another way to approach the question is through introspection, by examining ourselves and assigning to animals some of the different states that we recognize in our own minds. Stephen Toulmin, for example, distinguishes between sentience, attention, and articulation. Each one of these states can be either conscious or unconscious. A sleeping subject is only unconsciously sentient, while a wakeful subject receives external stimuli through the senses and experiences conscious sentience. We employ conscious attention when driving and taking mental note of what we see on the road, but if we are thinking about something else or talking to someone, we are driving on "automatic pilot." That is unconscious attention. We can also use the latter phrase to describe how, according to Steven Mithen, archaic humans (ones not of our species) made tools. Within Toulmin's schema, conscious articulation is a behavior that obeys well-established plans that can themselves be described to another individual. Unconscious articulation would be an activity with no clear motive.

It is difficult to recognize anything more than conscious sentience, or unconscious attention at the most, in animals or even in humans when we are small children, before learning to speak. Animals can neither make long-term plans nor observe their own behavior. These are two of the components of human consciousness. I don't doubt that in addition to sentience, animals have knowledge and desires, since they do know and want, but they do not seem to be able to analyze their own knowledge or desires. They do not know what they know or what they want because they lack an inward looking "third eye." Human consciousness is introspective, so we are conscious of having consciousness and we spend time philosophizing about it, asking questions like, "Where did it come from?" and "How did we get it?" But we could not be more alone in the world than when we engage in such philosophical reflection!

My only doubt, and it is a big one, concerns the chimpanzees, who seem very close to a limited consciousness of themselves, possibly like that of our common ancestors five or at the most six million years ago. A series of experiments initiated by Gordon Gallup seems to have demonstrated that chimpanzees recognize themselves in a mirror, something that no other animal can do except for orangutans and some, but probably not most, gorillas. In these experiments, the forehead and ears of an anesthetized chimpanzee are marked with paint, and the animal is later placed in front of a mirror. The chimpanzee raises its hand to touch the painted area, which it can see reflected but not directly. This may indicate that the animal knows that it is him or herself reflected in the mirror. There may be other, less obvious, interpretations, though. Euan Macphail, for example, speculates that the chimpanzee is only using the mirror to guide its hand towards a blotch of paint on the body of a chimpanzee that he or she sees there, without knowing that it is itself.

But let us assume that Gordon Gallup's interpretation is correct. This is more than a curiosity. The implication is that chimpanzees are self-conscious, and presumably the first hominids were as well. They had egos. Some think that self-consciousness could have evolved as a useful mechanism in the development of social behavior. To speculate on the possible actions of another individual, so as to prepare a response to those actions, it is most useful to imagine being in his or her place and to ask oneself, "What would I do in that situation?" If chimpanzees can do this, that means that they are able to represent the minds of others in their own minds, something truly phenomenal.

The longstanding discussion on whether chimpanzees have this or that manifestation of consciousness shows no sign of abating. This alone is enough to convince me that they are a borderline species between the instinctive animal and the fully conscious human. People often ask me why chimpanzees have not passed decisively over the threshold and into consciousness, why they are still apes, and why they haven't "evolved more." The answer is that for one thing, only after millions of years did our ancestors take that step, when *Homo ergaster* emerged onto the scene. Then again, encephalization is only one of the possible paths of evolution. The chimpanzees have followed a different path. They have continued to evolve but not towards greater encephalization.

Descartes versus Wittgenstein

Up to this point we have tried to learn something about the minds of animals, if indeed they have minds. In reality, the only thing we know for sure is what their minds are not. They are not like ours. The animals' lack of language makes it impossible for us to read their minds, which are therefore completely opaque to us. People's minds seem transparent to us though, and we read them all the time. That is how we know how to behave in the presence of others. We interact and make various kinds of deals with them. Sometimes those deals don't work out so well for us. After all, the only mind we really know fully is our own.

René Descartes based his philosophy on this human knowledge of the existence of our own minds. All else can be questioned, but his *Cogito, ergo sum,* "I think, therefore I am," gives us something to hold onto, something we can be sure of. And according to Descartes, other truths can be deduced on the basis of this certainty, the existence of God and of the world, in that order. But, he said, there are two kinds of worlds, the internal and the external. The essence of the internal world is thought, or consciousness. To Descartes, the human body was an animated machine like the body of any other animal. But an immortal soul was coupled with this particular machine, just as a captain forms a unit with his ship. Descartes did not believe that animals had souls, and it was said that he used this argument to try to justify his scientific experimentation on live animals.

The important point here about Descartes' thinking is that by assuming mind/body dualism, he arrived at an interpretation of consciousness that is essentially identical to the paradigm of the homunculus. The homunculus is a little person, an audience of one at a theater in the human brain where all the objects and events in the outside world are reproduced on stage. In a modern version of this paradigm, the little person would be watching television. If we throw Sigmund Freud's version of things into the pot, there would be an additional homunculus, the subconscious, locked in an alcove of the theater. But animals have no "animalunculus." They don't have a little theater in their mind where objects and events are reproduced, but even if they did, they have no internal spectator to attend the performance. Thus animals lack both self-awareness and perceptive awareness, or consciousness. They are no more than biological machines.

I deeply apologize to all cat- and dog-lovers for saying so, but these animals do not seem to be self-aware or able to represent the world internally, and they only possibly have perceptive consciousness.

But all or most of the so-called "superior" primates, that is the simians or anthropoids, the animals with brains most similar to ours, may have at least visual consciousness. It is impressive to learn that more than fifty percent of these animals' cerebral neurons and total cerebral volume is dedicated to the very complex task of processing of visual information. The largest computers, seemingly so intelligent when they perform advanced calculations, have very little capacity to recognize images or to discriminate between them. A simian, on the other hand, does not have the luxury of confusing immature and ripe fruit. And we have already discussed the possibility that our brothers, the chimpanzees, are self-conscious.

None of the above justifies the torture of animals, even for research purposes, unless those purposes are critically important to saving human lives, clearly a greater good. Since the time of

Descartes, there has been much discussion on whether animals have at least "sentient consciousness," whether they feel pain. When a dog withdraws its paw from a fire and yelps, does it really feel pain, or is it simply preprogrammed to move its body away from what is burning it and to advise others, especially its relatives, of the danger? These behaviors clearly improve the biological position of the individual, increasing its chances to impart its genes to another generation. There is no way to directly ascertain what sensations animals feel *inside* or if they feel anything at all. Within Descartes' mechanistic framework, the answer is negative, but the logic of evolution suggests a positive answer, because it would seem more adaptive to feel pain than not to feel pain. Pain is an intense and exigent subjective experience that forces us to focus our concentration on the most urgent circumstance of the moment, dropping everything else, so to speak. As someone once said, if you have a truly painful tooth, you can lose your whole soul in that one little cavity. This mechanism, this compelling sensation, seems to provide an excellent stimulus for reacting to danger and also for learning from experience, leaving one with an unforgettable memory associated with whatever circumstance has caused pain.

But we can attribute sentient consciousness, or conscious sentience in Moutin's categories described above, to many animals, especially the mammals. The philosopher Jesús Mosterín does so in his recent book *¡Vivan los Animales!* (*Long Live the Animals!*). Another question, whose answer is even more elusive, concerns the very nature of pain as animals experience it. Beyond their sentience, do they feel anguish, fear, frustration, or depression? Do they suffer? And also, do they experience hopefulness or happiness?

Mammals, and especially primates, employ facial expressions that in humans would unambiguously communicate suffering or happiness, but we are still plagued by doubt. After all, when a dog approaches its human friend, wagging its tail vigorously, is it really experiencing glee, or is it simply that such friendly behavior towards a dominant animal was selected in its ancestors the wolves because it had material benefits? I prefer to think the former, but I can't help remembering the words of Descartes, who thought it incongruous when certain individuals opined that their beloved canines had "souls" yet enthusiatically supped on mutton stew.

Descartes was a French philosopher who lived from 1596 to 1650, but the Athenian philosopher Plato (427-347 B.C.) had long before established a partial antecedent to his conception of the human mind. Plato said that before finding a corporal home, the soul lived with the gods in a paradise of pure ideas. In this material world, there are no pure and constant ideas, only objects, variable by nature. We are able to manipulate concepts because the objects that we see and touch are shadows and reminders of the pure ideas that we knew in that previous world. According to Plato, this is the only explanation for our capacity to establish categories that in fact have no observable existence in the world we call real. No one has ever seen the *Tree* as an idea, only a lot of large plants that we group together under that name. And of course no one has ever cast a mortal eye upon the "ideals" of Justice, Love, Beauty, or Wisdom.

To Descartes the soul produces ideas, while to Plato it only remembers them. In either case, the result is that upon thinking, the mind (or soul) manipulates ideas, either newly produced or remembered, and expresses them through words when there is the desire to communicate them to another mind. Language is but the instrument that makes it possible for ideas to travel from one mind to another. It is their vehicle. With this understanding, we can distinguish between the signified (idea) and the signifier (the conveyer of the signified). Words are signifiers and the signified is the concept that one wishes to express. A person, or perhaps I should say a homunculus, who speaks more than one language, can choose between different words to use as vehicles for the concepts that he or she wishes to convey.

Many different philosophical arguments may be employed to support the central concept of mind/body dualism. First of all, nobody believes that the loss of a leg or any other corporal mutilation, or even the loss of the language faculty, would mean more than the most minimal diminution of his or her essential personhood. Secondly, leading linguists and in particular the most famous linguist of all, Noam Chomsky, believe that we are born with a an internal device specifically designed for the acquisition of language. This device would be at the service of the mind and its expressive function, similar to how a computer peripheral interacts with the central processing unit. Thirdly, when damage is done to either of the two areas in the left hemisphere of the brain, called Broca's area and Wernicke's area, language capacity is negatively impacted. Damage to Broca's area makes speech very difficult and damage to Wernicke's area sorely affects language comprehension, although hearing acuity is unchanged. The patient hears perfectly well, but words seem to have no meaning. The localization of language faculties in specific areas of the brain bears out the idea that language is a peripheral capacity, an accessory if you will, which could correspond to certain neurological structures constituting a "language organ." The mind itself is not localized in any specific region of the brain. On the contrary, it affects the general functioning of all the brain's different parts.

Jerry Fodor, an influential contemporary psychologist, proposes a division of the mind into perception and cognition. Perception would be accomplished through a series of innate and mutually independent modules, which are more or less prepared for use at birth. Following Chomsky, Fodor includes language in this category. Cognition, on the other hand, would take place in a central system that carries out the mental activities commonly called *thought*. This central system is inaccessible to researchers and remains a mystery.

The computer analogy provides a vigorously modern and non-religious version of the Cartesian conception of mind. The language capacity can be understood to reside in a preprogrammed module contained within an actual physical structure that snaps into place in some section of the computer's innards and is prepared to process input from the time of birth. It needs to be filled with the data that constitutes the lexicon, or vocabulary, of a language, but syntax, the fundamental rules of human language, is hardwired in the device like the integrated circuits stamped on a computer's circuit boards. If this is true, then some day we will come to know the universal grammar that is common to all languages. Unfortunately, progress up to this point leaves much to be desired. Since its function is to enable our communication with other humans, we can group the language acquisition device, or the language organ, in the general category of relational modules along with those other peripheral devices, the sensory organs.

The mind, on the other hand, does not correspond directly to any physical structure, because it is the basic program of the computer, the group of instructions that makes it function and carry out computations. The lowest level programming language of a computer is its machine code, a binary system that functions entirely on the basis of two alternatives, conventionally represented as 0 and 1, and functionally equivalent to *on* and *off*. On top of this binary code, which is the only language understood by the computer itself, is the operating system. The OS in turn supports applications for the processing of words and images, navigation through calculations and the Internet. Thanks to these applications, we can communicate with our computers.

Everyone knows how to speak from a very early age, but nobody learns physics or mathematics naturally. On the contrary, considerable effort and a degree of maturity are needed in order to study these areas of knowledge successfully. Following the digital computer analogy, we can say that the computer comes preprogrammed with the underlying rules of grammar. They are physically recorded (hardwired) in certain circuits of the machine. But science and humanities software may be installed or not, according to the preferences of the operator. In any case, they are information, not circuits, and will be stored in another area of the computer. So in this analogy, mind and language are two different things. The invisible and ethereal nature of programming and its quasipure attributes confer upon it a strange kind of spiritual quality that makes the computer science analogy irresistible to some people. How seductive: science and magic rolled into one, a new religion for a new century.

There are some who aspire to space travel in the form of bytes. Personally, I am too much at home in my body to want to see myself burned onto a compact disk. It would not be so simple, anyway. Although the most elementary chess program can beat me at the game by means of computation, I do not detect the slightest sign of reflection on its part. I am not even impressed by Deep Blue, the machine said to have beaten chess champion Garry Kasparov. I truly think that an ant is more talented. Will a conscious machine be produced some day? Will it have human emotions, too? This is an old dream, or nightmare, of humankind. Some say that it will soon come true, but I don't believe it.

There is a radically different way to address mind/body duality, with its roots in the thought of the Viennese philosopher

Ludwig Wittgenstein (1889-1951) and his followers, especially Gilbert Ryle. They simply deny the existence of the individual mind, which they consider an unnecessary myth, the reification (treating an abstraction as a concrete thing) of something that is purely conceptual. Given that we act consciously, they say, we erroneously believe that a real entity that we call consciousness exists from the time of birth and is the source of this behavior.

But if the mind does not exist, who or what is carrying out all our mental operations? If there are no homunculi inside our heads, who or what perceives, knows, recognizes, decides, remembers, and speaks? The answer is nobody, or in a sense everybody who belongs to a given community. The mind, in their view, is not a private entity or a personal resource of any individual, but something shared socially within that community.

It is we adults who provide the input to structure the minds of small children. We lead them to believe *x* or *y*. To do this, we employ the tool of language and the very effective technique of determining what children are to learn by directing their attention towards what interests us. Thus do we teach children what it means to be human. It is true that the offspring of other mammals, particularly the social species, learn from their elders by observation and imitation. They are also corrected when their behavior is unacceptable. But there is no animal behavior similar to the instructional methodology practiced by humans with their children. Human knowledge is definitely acquired through social interaction. Only the capacity to acquire it is innate.

According to this school, we believe in the individual and innate mind because we take things like deciding, understanding, and perceiving to be processes or operations, when in fact they are results. The word *mind* refers more to a kind of behavior than to a real entity. If an operation is performed, then there must be an agent performing it. On the other hand, if no operation is being performed, the concept of agency is superfluous. When we say that a tree has been seen by someone, we are describing a result, not a process. An object is perceived to be a tree when we attach a label to it, the word *tree*, which our language community agrees upon as the name for a certain kind of vegetation. However, it is not always clear if a plant is a tree or a bush, because the boundary between large- and medium-sized flora is not unequivocal. However, it would be crystal clear if it were a boundary between two of Plato's pure ideas instead of just two social conventions. A child may err and call a large fern a tree, and he or she will be corrected. The name of a thing and the meaning of a word have both been understood correctly by an individual when the social collectivity concurs with his or her usage of word *x* for thing *y*.

William Noble and Iain Davidson, researchers into the origin of language and mind in human evolution, are thinking along these lines and in the tradition of Wittgenstein. They believe that since mind, or consciousness, does not exist without language, we must believe that the two phenomena originated at the same time, which they say coincided with the appearance of our species. In their view, no other hominid, including the Neanderthals and our premodern ancestors, had consciousness. The classical perspective on mind, on the other hand, does allow for the possibility of a nonverbal or mute consciousness before the appearance of language in human evolution, given that consciousness and language are separate phenomena and up to a point independent.

Noble and Davidson go so far as to assert that the individual and innate mind is a product of Western philosophy, that we have this idea only because it occurred to Descartes to propose it. But

I disagree. I think that since the concept of mind is universal to all human cultures, it must have a strong innate component. I think it stems in some unknown way from the nature and organization of our cerebral cortex. The relationship between consciousness and language is an even thornier problem. Still, I do agree with Noble and Davidson on one thing. Given that consciousness is such a slippery concept and so difficult to pin down, why not concentrate first on language, which can be described much more easily? The definition of language that they offer is very simple: any system of communication by means of symbols.

In order to determine what exactly constitutes a symbol, we can consult Charles Pierce's classification of signs. According to this late nineteenth-century philosopher, a sign is simply a thing that represents another thing. Signs are then divided into three categories: icons, indices, and symbols. Icons are associated with their referents by a similarity in appearance. The most obvious example is a drawing, which may be more or less detailed, but which must necessarily share some visual characteristic with the object it purports to represent. A map may also be considered an icon, and in the auditory sphere, an onomatopoeia is an icon, because it imitates the sound of its referent. Indices do not resemble their referents, but they are causally related to them. They are produced by the object they represent and are limited by its characteristics. Smoke is an index of fire. Sherlock Holmes sought indices in the identifying characteristics and circumstantial evidence that he used to solve his cases. Symbols, on the other hand, are completely arbitrary. There is no need for them to resemble their referents or to be conceptually related to them in any way. Spoken and written words are symbols, as are the signs of codified gestural languages. While symbols are intrinsically arbitrary, they become meaningful by means of social convention.

The Morse Code of telegraphy is quintessentially symbolic. By means of long and short electrical impulses that we call dashes and dots, letters and words are transmitted. In order to understand a message, one must know two languages, Morse and English, or Morse and any other language written with the Roman alphabet. We also communicate using icons, on some traffic signs for example. Other traffic signs are completely arbitrary, i.e. purely symbolic. Icons can sometimes do double duty as symbols. A heart shape, for example, is an icon for one of our internal organs but also a symbol of love. A symbol can sometimes be a set of disparate objects in association with each other. In the West, for example, a robed and blindfolded woman holding a balance scale means justice. While icons and indices are universally understood, arbitrary symbols only make sense within a language community that collectively subscribes to a convention, a tacit agreement, for example to express the idea or the ideal of justice by means of a blindfolded woman with a balance. Another symbolic convention would be to signal a state of mourning by wearing black.

Like Pavlov's dogs, domestic animals can be trained to react in a consistent and predictable manner in the presence of signs produced by humans, but that does not mean that they understand them. It makes no difference to them if they are icons, indices, or symbols. They simply establish an association through the experience of positive or negative conditioning. A dog may sit when a person so commands, but it would be absurd to conclude that it understands human language. In any case, it is a fact that animals do not communicate by means of symbols, and they cannot understand the meaning of even the simplest icons, much less employ them in communication with others of their kind.

I would like to suggest an example particularly relevant to the lives of our ancestors. According to the linguist Georges Mounin, indices are "…phenomena that are immediately perceptible that inform us about other phenomena that are not." All predators localize their prey, and all prey animals seek to localize their predators. Using their visual, olfactory, and auditory senses, they identify each other by shape, smell, or sound. We can say that an animal produces sounds that are indices of its presence, but that its scent and shape are two of its perceptible attributes. Since the scent of an animal lingers after its body has passed, a predator can follow its trail by smell. But no animal can recognize and follow the tracks of another. There is no Sherlock Holmes in the animal kingdom.

But again the case of the chimpanzees raises doubts. Some individual chimps have been "educated" in the use of language. For physiological reasons, they are not able to speak, so they have been shown how to use a gestural language of deaf-mute humans or to use a special computer keyboard. The chimpanzees learn how to use individual and paired words correctly, but rarely manage to handle three-word phrases. Yet we must ask, to what extent they understand the words they use? Have they been taught or just trained? Some chimpanzees have been able to master vocabularies of more than one hundred fifty words or signs. By "master" I mean that they use the words in appropriate contexts and respond appropriately when the words are used to make requests of them. Whether or not they capture the semantic content, that is to say the meaning, of the words is worth discussing, but they are not at all adept at ordering words in phrases or applying what we call grammar. Of course the latter is equally important for effective communication. It is very difficult for chimpanzees to understand that "Put the cup on the plate" is not the same as "Put the plate on the cup." A small child, in contrast, quickly masters fundamental grammatical rules such as the customary English word order of subject-verb-object.

I was pretty skeptical about even the elementary symbolic and linguistic capacity of chimpanzees until recently, when I saw a film biography of the chimpanzee Washoe. This animal, although I hesitate to use that word to refer to such a famous creature, impressed the entire world some years ago with her remarkable linguistic skills. The husband-and-wife research team of Allen and Beatrix Gardner conducted pioneering psychology studies with Washoe that produced extraordinary results. Late in her long and often sad life, Washoe had a baby daughter that fell ill. Lab workers removed it for treatment, but it died and was not returned to her. Subsequently, Washoe would relentlessly repeat just two signs whenever the research psychologist working with her came to her cage: "Bring baby, bring baby."

Darwin versus Wallace

There is as yet no model to explain how consciousness and language appeared in evolution, either separately or conjointly. When I say "how," I mean to ask by what sort of mechanism were we enabled to differ so radically from all other creatures.

Charles Darwin and Alfred Russell Wallace, the two codiscoverers of the theory of evolution by means of natural selection, sharply disagreed on this point. To Darwin, the evolution of the human mind did not substantially differ from the evolution of the body. A long evolutionary road separated humankind from the apes. Enormous changes had been made by means of small, incremental steps over an extended period of time. He made this clear in his only reference to human origins, near the end of his most famous work, *The Origin of Species*: "In the future I see open fields for far more important researches. Psychology will be securely based on the foundation already well laid by Mr. Herbert Spencer, that of the necessary acquirement of each mental power and capacity by gradation. Much light will be thrown on the origin of man and his history."

Wallace simply could not accept the idea that humankind's elevated moral and intellectual faculties had been produced through gradual evolution. They were too radically distinct to represent an accumulation of incremental changes. Instead he saw us becoming human in one great qualitative leap, a leap that had to have been supernatural.

Ian Tattersall, a prominent paleoanthropologist and a good friend of mine, believes that both Darwin and Wallace were onto something. Tattersall sees what is often described as the *emergence* of human cognitive capacities as a good example of what are known in systems theory as *emergent properties*. The functioning of a system and the set of properties that characterize it are conditioned by the elements that make up that system and the manner in which those elements interact among themselves. Any innovative rearrangement of its elements has the potential to endow the system with a revolutionary new property, completely and radically different from all previously existing properties. Though the result may seem miraculous, it is not. This is science, not magic.

In applying this analysis to a biological system such as a living organism, we can say that each of the organism's recognizable characteristics is one of its constituent elements. The word *adaptation* has traditionally been used to describe any characteristic with a recognizable function. From time to time, it has been demonstrated that a characteristic was related to a certain function when it appeared, but over the evolutionary history of a particular group it took on an entirely different function. These characteristics are called *preadaptations*. The history of feathers provides an example. Feathers isolate and retain body heat very well. They seem to have developed as a layer of thermal insulation on the bodies of a certain group of dinosaurs. But feathers came to be used by birds, a subgroup of the plumed dinosaurs, to enable or enhance flight.

Since the term preadaptation may suggest some kind of pre-destination, we have come to distinguish instead between an *aptation*, any characteristic related to a particular function; an *adaptation*, which now refers only to a characteristic that has not changed its function since its origin; and an *exaptation*, a characteristic which has done so. The last term is conceptually equivalent to the formerly employed *preadaptation*. According to Tattersall, our large brains and our vocal apparatus, physiologically able to produce articulate speech, are exaptations. They emerged in functional contexts distinct from their present ones, cognition and language, respectively. Once acquired, they were nevertheless unable to produce either cognition or language until new neurological pathways established the necessary connections between them. In other words, when the elements of the system took up a new configuration, evolution pulled two rabbits out of its hat, the revolutionary emergent properties of mind and its inseparable companion, language. Tattersall's theory is very appealing. Still, it is not easy to understand why our ancestors' and the Neanderthals' brains would have grown so large when we know how costly it is to supply the brain with the disproportionate amount of energy it requires, and if as Tattersall thinks, these archaic humans operated on the basis of instinct rather than cognition. Then too, why would the articulatory apparatus develop before the existence of symbolic language?

Steven Mithen has recently proposed a theory that takes a very different approach, although he concurs with Tattersall that the human mind appeared or emerged rather suddenly due to a reorganization of previously existing elements. This sudden evolutionary development would certainly have surprised an outside observer, had there been one.

In addition to the classical theory that sees mind as a separate entity from body and in particular to each individual from birth, and the opposite theory that denies the existence of the individual mind in favor of collective consciousness, there is a third possibility. It is the theory of multiple intelligences. Steven Mithen bases his ideas on the work of Jerry Fodor, Howard Gardner, and the evolutionary psychologists. Mithen's evolutionary model has several stages. In the first phase, the period of the Australopithecines, there would have been a general intelligence competent to resolve routine problems, a social intelligence used in the formulation of intragroup relations, and a natural sciences module specializing in the relationship of the individual with his or her ecological environment. Consciousness of self would have developed, as mentioned earlier, on the basis of social intelligence, and would not have developed beyond its utility in that context. Mithen compares this first phase to the condition of modern chimpanzees.

At a later stage of human evolution, with the appearance of the first members of the genus *Homo*, a new technological intelligence would have emerged, enabling them to fabricate stone tools. Making tools may have been somewhat difficult for them, but these people would have been able to perform the requisite tasks without considering what they were doing. The number of very complex operations that any one of us does automatically every day is truly impressive, and we are certainly not conscious of everything that passes through our minds.

The first rudiments of language would have been developing at the same time, but exclusively within the domain of social intelligence. Later humans, like the Neanderthals and our premodern ancestors, would have further developed their general, social, ecological, and technical intelligences, but each intelligence would have remained autonomous unto itself and language would have been limited to the communication of social information.

Finally, with the appearance of our species, the barriers between separate intelligences would have broken down, and consciousness and language would have been extended to all domains.

Although I sympathize with the idea of consciousness expanding over time, I see three problems with Mithen's theory. To begin with, it is difficult to see how advanced technological skills or broad ecological knowledge could have been achieved in an automatic and unconscious, i.e. instinctive, mode.

In the second place, it is almost impossible to accept the idea of language restricted to social relations. The essence of language is communication by means of symbols. I can understand the idea that a species may have a greater or lesser capacity to manipulate symbols, but not that it would have the capacity to manipulate only one particular category of symbols, if we can even say that there are distinct categories of symbols.

Chimpanzees unconsciously produce vocalizations that usually express an emotional state, be it anger directed at a rival or a stranger, joy at finding a fruit-laden tree, or fear in the presence of a dangerous predator. They are very physically expressive, gesticulating as they vocalize. As we have seen, chimpanzees also possess an impressive capacity to process visual information, including body language. This information, whether anger, or the presence of figs, or a leopard, is extremely useful to other members of the group. They are interested both in the emotional state of their comrade and in the cause of that state. But at the moment when an ancestor of ours was intelligent enough to understand the effect that his or her vocalizations and gestures were having on others, he or she understood its meaning and language was born. His or her vocalizations and gestures automatically became symbols that could be modified and manipulated in order to transmit information, even false information, at will. For this to have happened, the hominids who "discovered" language also had to read the minds of their fellows, so they had a "theory of mind" and were unambiguously self-conscious. Once language was invented, any kind of information could be represented symbolically. It is not even really important to know if visual language based on gestures preceded oral language based on sounds or if the two developed in tandem.

Finally, Mithen has proposed a process of humanization that, as in Tattersall's model, could accommodate gradual change rather than a sudden transformation coinciding with the appearance of our species.

We will see in the following section if the models proposed by Tattersall and Mithen, which conciliate Darwin's and Wallace's points of view, accord with the available data on human evolution, or whether, as I fear, we will be obliged to choose between Darwin and Wallace.

In Principio Erat Verbum— In the Beginning Was the Word

After a long digression on the nature of mind and language, the time has come to examine the archeological and paleontological record for signs of their development. But first let us consider what the fossil record has taught us about the evolutionary history of Neanderthals and modern humans. If we are to compare the minds of the two species, then human evolution can be divided into two long stages. The first stage is a shared story that began with the first hominid and continued right up to the day when humans arrived in Europe. At first, the European populations, represented by the fossils of *Homo antecesor* at Gran Dolina, did not differ from those in Africa or western Asia, while a distinct human species, *Homo erectus*, lived in east Asia. But over a prolonged period of isolation, the Europeans began to differentiate themselves. Even before the time of Sima de los Huesos, some 350,000 to 400,000 years ago, they had developed some Neanderthal-like characteristics. Nevertheless, the European and African populations still shared many primitive characteristics inherited from their last common ancestor. Some time later, about 100,000 years ago, the European populations had evolved into Neanderthals, while our ancestors, somewhat archaic but clearly modern, lived in Africa and Palestine.

We can identify two moments of accelerated change in the expansion of the human brain. One corresponds to the time of the first humans, *Homo habilis* and especially *Homo ergaster*, in Africa when brain volume doubled. The second alteration of the evolutionary rhythm took place independently in both Europe and Africa about 350,000 to 400,000 years ago and produced the large brains of the Neanderthals and of modern humans. The Sima de los Huesos fossils represent the very moment when brain size really began to take off in Europe. We still know little about the last ancestor common to us and to the Neanderthals, the species represented at Gran Dolina, but we can suppose that the evolutionary state of its brain was intermediate in relation to these two moments. So the mental capacities that both we and the Neanderthals developed must have been either a shared inheritance of that remote common ancestor or a case of independent but parallel evolution. In the following paragraphs, we will review the possible manifestations of those capacities, and we will discuss how they were acquired and what they mean.

I will begin with cerebral morphology. As I have mentioned before, the Neanderthals probably achieved an intelligence very close to our own. A study of evolutionary tendencies in the two lines clearly points to parallel encephalization. The Cro-Magnons do not appear to have had any demonstrable advantage as measured by the ratio of cerebral volume to body weight. But in addition to overall cerebral volume, we must consider the proportional volume of the brain's constituent parts. Our brains are not simply larger versions of chimpanzee brains. With the growth in our brain size, certain areas of the cerebral cortex were reduced in proportion to others. One example is the primary visual cortex, located at the posterior pole of the brain, inside the occipital lobe. Other parts of the brain became relatively larger, notably the prefrontal associative cortex in the frontal lobe, the most anterior portion of the brain. Higher, specifically human, functions are attributed to the prefrontal cortex. Thanks to the capacities of this region, we can store information and keep it "on line" for retrieval when circumstances make it desirable to do so. This makes it possible to remember the extended gestural sequences necessary to complete a complex task such as producing a stone tool or playing the piano.

The prefrontal cortex is also connected with certain structures near the center of the brain that, taken together, are called the limbic system. The limbic system plays a key role in our emotional lives. An injury to the prefrontal area or its surgical amputation, called a lobotomy, can dramatically transform personality. It seems to be the seat of attention, of self-consciousness, of the ability to make plans for the future, and of the motivation to carry them out. Above all, it endows us with our imagination and creativity. In addition to the prefrontal region, other associative areas in the parietal, temporal, and occipital lobes are proportionally larger in the modern brain than in the brains of our ancestors. As far as I know, no one has demonstrated that the Neanderthals substantially differed from us in any of these aspects.

The asymmetry of our brains is notable. They can be divided into two halves called hemispheres; the left hemisphere of right-handed people projects occipitally, or more towards the rear, than the right hemisphere, while the right hemisphere projects frontally, or more towards the front, than the left. Neither this marked asymmetry nor such strongly marked handedness is found among other primates. Chimpanzees, for example, are essentially ambidextrous. This curious difference may be of some importance, since certain cerebral functions seem to be more characteristic of one side of the brain than the other, and the cerebral activity that controls the use of language is among them.

As I have noted, both Broca's area and Wernicke's area are located in the left cerebral hemisphere. The classical Broca's area, in the frontal lobe, seems to play a role primarily in the motor sequences necessary for speech production and for the control of other activities with little relation to language. Wernicke's area is located near the juncture of the temporal, parietal, and occipital lobes. It is very important to the comprehension of language and of symbols in general. Modern brain mapping and PET, Positron Emission Tomography, have demonstrated that many other areas of the brain are involved in the production and comprehension of language. Since they too are principally located in the left hemisphere, there is a strong temptation to think that the existence of language is dependent on brain asymmetry and body lateralization.

Working with the paleoanthropologist Ana Gracia, I have observed cerebral asymmetries similar to those of modern right-handers in fossil crania from Sima de los Huesos. José María Bermúdez de Castro and colleagues have demonstrated that those individuals used their right hands preferentially. Surprisingly, this did not involve an examination of their bones, but of their teeth! The Neanderthals would hold one end of a piece of meat in their mouths while cutting it with a handheld stone tool. At times the knife would slip and leave marks on the anterior face of the upper and lower incisors. These marks tell us which hand was being used to cut. If the mark runs from upper left to lower right, it was the right hand. If the mark is from the upper right to the lower left, the left hand was the cutting hand. In the available sample, the majority were right-handed, as most of us are today.

Philip Tobias and Dean Falk have provided other evidence for the existence of language among the first representatives of the genus *Homo*. They believe that they have found the impression of a well-developed Broca's area on the interior surface of a 1.8-million-year-old cranium. This phenomenon is never observed in other primates. Rick Kay, another researcher, has observed that the diameter of the two hypoglossal canals in the Neanderthals was as large as ours in proportion to the dimensions of the oral cavity. The hypoglossal canals are at the base of the cranium below the two occipital condyles, which articulate the skull and the atlas, the first cervical vertebra. The two hypoglossal nerves, crucial to the fine motor control of the tongue, pass through the canals. The fact that the diameter of the hypoglossal canals was large in relation to the size of the oral cavity means that they were also large in relation to the size of the tongue. This we must deduce, since the tongue itself does not fossilize. It could also mean that the hypoglossal nerves were large and contained many nerve fibers, making possible the production of a wide and well-differentiated set of speech sounds.

Continuing in our search for aspects of language capacity, we can turn from the brain and nervous system to fossils representing the phonetic apparatus, where speech is physically produced. Phonation itself occurs at the vocal cords of the larynx, a set of folds that open and close to permit or inhibit air expelled from the lungs to pass between them. Up to this point, humans present no innovation, since all primates produce vocalizations. The features unique to humans are a little higher, in the airways above the larynx called the supralaryngeal airways or the supralaryngeal vocal tract.

The oral and nasal cavities are separated by the palate. The palate itself is divided into the hard palate, which is the anterior bony section, and the soft palate behind it, also known as the velum, which has no bony support. The velum in turn ends posteriorly with the uvula, the pendent fleshy lobe at the back of the mouth. Only mammals have these palatal structures between the oral and nasal cavities. It is an adaptation that separates the two cavities and makes it possible to breathe through the nose when the mouth is completely obstructed by food and pulmonary air cannot pass through it. Zoologists call the roof of the mouth the secondary palate, because it is a kind of false roof located below the original, or primary, palate. Curiously enough, crocodiles have developed a secondary palate in an evolutionary convergence with mammals. It has similar functions.

The pharynx is behind the oral and nasal cavities, in the space between the end of the palate and the spinal column. It descends vertically to the larynx and the esophagus, the former situated in front of the latter. This makes for a two-part supralaryngeal airway, with a vertical and a horizontal section. In all mammals except for adult humans, the oral and nasal cavities and the pharynx are structured horizontally from front to back, or sagitally, in technical language. In other words, the palate is large and is also relatively distant from the spinal column, making the horizontal component of the supralaryngeal vocal tract quite long. But the larynx is very high and close to the mouth, so the vertical component of the vocal tract is very short.

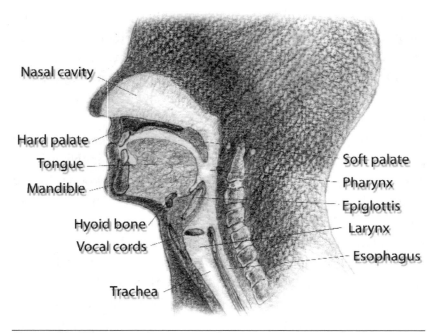

Figure 23: The upper digestive and respiratory tracts of our species.

In adult humans, the oral and nasal cavities and the pharynx are sagitally short, because the palate is shorter and also closer to the spinal cord. But since the larynx is lower, the pharynx is vertically longer. The respiratory and digestive tracts cross at the level of the pharynx, so food can get lodged at the entrance to the larynx or even enter the trachea instead of the esophagus. This presents a serious risk of death by choking. In nursing human infants and all other mammals, the larynx ascends like a periscope and seats itself at the base of the nasal cavity during swallowing. This allows liquid or solid food to pass across it without danger. Infants and other mammals can nurse or drink and breathe at the same time. At that time, the pharynx is temporarily behind the larynx. By the age of six or seven years, the human larynx has descended to its adult position, and the palate has moved closer, in relative terms, to the spinal column. All fossils of adult modern humans exhibit this morphology.

To compensate for the danger of choking to death, the adult human pharynx is a long vertical tube within which sound produced at the vocal cords can be modified, as in a wind instrument. As long as one is not swallowing, it can be used to produce the wide variety of sounds that characterize articulated human language. The principal method of sound differentiation is to change the configuration of the tongue and lips. While the tongues of all other mammals are thin and lie entirely within the mouth, our tongues are very thick and also serve as the anterior wall of the pharynx.

It is possible, although not at all easy, to reconstruct the supralaryngeal airway of fossil remains. The Australopithecines and the paranthropoids definitely had long horizontal segments and probably had short vertical segments, like chimpanzees for example, so we believe that they could not speak. To be more precise, we can simply state that the physiological apparatus that produces human speech as we know it today did not yet exist, though we do not know if they had developed a certain degree of control over the acoustic or gestural expression of their emotions.

With the appearance of modern humans, the horizontal portion of the supralaryngeal airway became shorter. This is the portion that extends from the front teeth to the spinal column. Two processes had brought about this change. The chewing mechanism was reduced in size, so the palate was shortened, and it also moved closer to the spine. In our species, the vertical and horizontal portions of the vocal tract are similar in length.

Jeffrey Laitman, a well-known expert on the origin of language, has stated that the backward movement of the palate in modern humans necessarily produced an arched skull base, so that an examination of fossil cranial bases will reveal which species could have had vocal apparatuses adequate for speech production. Ignacio Martínez and I are not so sure that there is a clear relationship between the arched cranial base and laryngeal position.

The Neanderthals had a somewhat reduced chewing apparatus as well, but the end of their palate was still considerably farther from the spine than that of the Cro-Magnons. In this particular morphology, the Neanderthals were like overgrown Cro-Magnon babies. Does this mean that they could not speak as we do? I think, along with Ignacio Martínez, that the answer has to be yes and no. On the one hand, the larynx had probably descended to a greater or lesser extent, producing a vertical tube within which to modulate the sounds of vocalization. At the Kebara site in Israel, a 60,000-year-old Neanderthal skeleton has been found with a very modern hyoid bone, the bone that supports the larynx. Although a morphologically modern hyoid does not necessarily mean that the larynx had descended, it makes one wonder. If this was the case, then Neanderthals would have been able to produce a wide range of sounds, but not exactly like ours, because the horizontal segment of their vocal tract was still primitive. In other words, we have a phonetic apparatus in which the horizontal and vertical segments are of similar length, while as we understand the Neanderthals, they had a vertical segment similar to or just a little bit shorter than ours, but a much longer horizontal segment. So theirs would have been a different instrument, producing different sounds. The words to the music may or may not have been the same, or they may not have been.

Some years ago, Philip Lieberman and his colleagues reconstructed a Neanderthal phonetic apparatus. They located the larynx lower than in our newborns, whose larynxes are very close to the mouth, but higher than those of our adults. This put the Neanderthal larynx in an intermediate position unlike that of any currently existing species at any age. Using a computer program, Philip Lieberman generated voices for the modern human infant and adult and for the Neanderthal. The computer-simulated voices were realistic for the two modern humans, allowing us to take the Neanderthal results seriously. The Neanderthal was able to produce a wide range of the sounds of modern human language, with the exception of the three vowels *i*, *u*, and *a*, and the two consonants, *k* and *g*. These symbols of the International Phonetic Alphabet represent the sounds that many Americans call a "long e," a "long u," and a "short o." The consonants are the "k sound" and the "hard g sound." This deficit does not seem to be dramatic. You could read this book perfectly well if it were rewritten without those letters, perhaps using others in their place. Some languages, like Arabic and Hebrew, are commonly written without any indication of vowel sounds at all, and they are are perfectly readable. But in spoken language, based on sounds rather than letters, the lack of these vowels and consonants would make a big difference. The vowels *i*, *u*, and *a* are called the cardinal vowels, and they are by far the most common vowels in human language. They are the only vowels used in Arabic, while in Spanish and Basque we have two more. Many other languages have even more vowels (remember, vowels sounds, not letters). English is among them, I suppose to frustrate those of us who speak it as a foreign language. The cardinal vowels, and in particular the *i* and the *u*, have the virtue of being the most clearly distinguishable vowels to the human ear. This has been demonstrated in many experiments. Their use in any language makes it much more intelligible, and thus easier for us to understand each other, particularly in the presence of ambient sounds such as noise or other conversation, a common circumstance. Thanks to the availability of the cardinal vowels, we can understand each other's speech without the need for total silence and absolute attention.

If the palate was further forward in the Neanderthals, their speech sounds would also have been more nasalized than ours because they would have had no mechanism to block the egressive airstream from entering the nasal cavity and exiting the nostrils. Unfortunately, more nasalized speech sounds are also harder to distinguish from each other. So Neanderthal linguistic capacity, the mental basis for communication by means of symbols, in this case words, may have been equal to our own, but their vocal tract would have prevented them from producing a set of sounds as distinct as ours.

Fossil Behavior

Using the material evidence that prehistorical humans left us during their lives, we may be able to find a sign, even if it is but a slight one, that they experienced consciousness. Behavior as such does not fossilize, but its results can. The tools that we humans have produced are among the most significant indicators that our cognitive capacity surpasses that of any other species. No other animal can even split a rock to produce a blade, much less produce a conceptually sophisticated instrument such as a biface or a Levallois point. The control of fire is another sign of superior mental abilities. We cannot even dream of a chimpanzee making a fire, whether it be by rubbing two sticks together or by striking one stone against another to throw a spark into a bed of kindling. Where there is controlled fire, there are humans. Humans are also the only animal we can observe to cry in mourning for its dead or to treat their lifeless bodies with respect.

All of these activities can be learned and performed by imitation and without recourse to language, but they also reveal a high level of consciousness. In fact, language itself is learned by imitation. For those like Noble and Davidson, who postulate the impossibility of consciousness, or mind, without language, the existence of consciousness among "archaic" (not modern) humans would imply the existence of language as well. So they try to demonstrate that no evidence of fire or burial is associated with any remains other than modern humans. They assert that bifaces were not produced intentionally and that Levallois technology never existed. But let us consider one thing at a time.

Ashes are often encountered at prehistoric sites, a sure sign of wood combustion. Ashes more than a million years old at the Swartkrans cave in South Africa have been attributed to human activity, possibly by members of the species *Homo ergaster*. It is possible, however, that the ashes were produced naturally outside the cave and later entered it in combination with earth or mud. It is also possible that underbrush may have grown into the cave mouth and then burned. In dry ecosystems, spontaneous fires are anything but extraordinary. In fact, you can count on them. Some plants are pyrophilic, or fire lovers. In Spain we have the example of the spotted rockroses. Fire not only stimulates their seeds to germinate better, but also eliminates those of their competitors not adapted to periodic burnoffs. After a fire, with surrounding vegetation eliminated, these "children of fire" find their place and spread exuberantly. Nor is it extraordinary for the heat of the sun, rather than humans, to provoke forest fires. The simple presence of carbon at a dig does not definitely prove that the humans who lived there made controlled use of fire.

At the approximately 500,000-year-old Level 10 of the famous Zhoukoudian site, for example, excavators before the Second World War found burnt bones, ashes, and other evidence of fire. At the time, this was interpreted as proof that *Homo erectus* had used fire as he colonized northern China, where he had to brave a much colder climate than either of his ancestors in Africa or his contemporaries in tropical Java. But more recent analysis has cast doubt on the intentionality of these fires. Is it possible that *Homo erectus* never controlled fire after all? It would not be extraordinary, if we consider that when Europeans arrived on Tasmania, they found people who did not know how to produce fire, although that island is not tropical but maritime temperate, often cool and damp.

Whatever some may say, Neanderthals did use fire in a planned and systematic way. The Kebara cave in Israel, where most of a 60,000-year-old Neanderthal skeleton has been found, contains the remains of fires clustered in groups marking a series of hearths. But we do not need to stray so far from home. At Abric Romaní in the province of Barcelona, my friend Eudald Carbonell has excavated many hearths where Neanderthals warmed themselves with controlled fire. Did Neanderthals in Catalonia pass their winter evenings seated around cozy flames and glowing coals? Can we imagine a group of humans, albeit Neanderthals, mutely sprawled around the hearth? We really must credit the use of fire with some role in the development of the human mind. Aren't we, like the rockroses, "children of fire?"

There is a discipline within paleontology called paleoichnology. Paleoichnologists examine the behavior of ancient organisms based not on the fossil remains of the creatures themselves, but on traces of their activities. Paleoichnologists are best known for studying dinosaur tracks, but they also study trace evidence of activities such as eating, the construction or digging of shelters, etc. I like to say that archeology is but a specialty within paleoichnology because it is concerned with the activities of those organisms of the past who just happen to have been our ancestors and close relatives. But seriously, the most abundant evidence in prehistoric archeology consists of stone tools. We should examine that evidence from a paleoichnological perspective in order to learn about the mental capacities of those who engaged in stone-based industries.

As I discussed earlier in this book, it does not seem that the practitioners of the first industries, the Oldowan, were seeking to produce any predetermined tool morphology. Their goal was a certain functionality, be it for cutting, mashing, or another purpose. But with the appearance of bifaces a million and a half years ago and their perfect symmetry on two or even three planes, many of us see the conscious, planned, and deliberate production of predetermined forms. To deny this requires an alternative explanation for the shape of these objects. Noble and Davidson's explanation is that the bifaces were simply what was left over when nuclei were repeatedly struck in order to produce chips. The primary tools being produced were the chips, while the nucleus was the raw material, and was transported from place to place to serve as such. The humans may have used the biface itself secondarily as a tool when it could produce no additional chips, but in no case was the biface the final desired product of an operative chain. According to this theory, the biface was produced without consciously trying. To me this is an outlandish claim that is not compatible with the facts, so I cannot subscribe to it. Needless to say I see even greater consciousness at work in the Levallois technology also employed by the Neanderthals and lest we forget, by our proto-Cro-Magnon ancestors some 100,000 years ago.

The last and most mysterious of the behaviors indicated by the fossil evidence is the burial of the dead. Neanderthal skeletons have been found buried in caves, and modern human skeletons have been found buried in caves and in the open. There is only one known case that suggests any funerary practice before the time of the Neanderthals and modern humans and that is at Sima de los Huesos. Although bodies were amassed one atop the other in a place chosen for that purpose at the Sima, they were not *buried*, a process defined as the digging of a grave and the placing of a body within it.

Since it cannot be denied that burial of the dead is purposeful and requires planning and consciousness, the only option for those who deny that human species other than our own had the capacity for such activity is to deny that they engaged in it. Actually, according to their arguments, these other species are not human. But the scenario that they propose is that the numerous nearly complete Neanderthal bodies that have been found in caves were not buried. They ended up there as a result of actions by other, nonhuman, biological, and/or geological agents. Anything will do, from floods that washed the bodies into the caves to lions and hyenas who carried them into their dens. They also provide some even more amusing explanations. In one case, an author suggested that the roof collapsed on a Neanderthal while he slept in a cave at Shanidar, Iraq. I'm not joking. These suggestions were published in the ninety-eighth year... of the twentieth century! Following this reasoning, Neanderthal fossils are found in caves because that is where all the rare circumstances coalesce to create false or rather *natural* burials, because all fossils are the result of a burial of some sort. But this is why, some say, no Neanderthal gravesite has ever been found in the open. There are, however, graves of modern humans almost 30,000 years old in the open in both Europe and Australia.

In the interest of brevity, I cannot offer a detailed analysis here of each and every case. Let it be understood that I interpret many Neanderthal fossil finds as evidence of deliberate and intentional human burial. Some among them are recovered from very recent excavations that cannot be disparaged with the convenient mantra that earlier generations of scientists worked with a lack of rigor and an excess of imagination. It is not even necessary to discuss the question of Neanderthal burial, not least because I have spent years arguing that their ancestors at the Sierra de Atapuerca engaged in funerary behavior 350,000 to 400,000 years ago as the find at Sima de los Hueso establishes.

In order to make Neanderthals appear less like us, some have attempted to deny the symbolic value of their burials by asserting that they were not motivated by religious sentiments but by their love and devotion for the dead. If the pain of loss was what motivated the Neanderthals to bury their loved ones, I offer them my personal condolences in their moment of grief. In my view, nothing could make them any more human than the idea of them weeping at "secular funerals."

Of course no one could doubt the existence of symbolic behavior in relation to the burial if it were proven that funeral rituals were practiced. The word "ritual" would need to be defined, but any time that an object is buried with the deceased it could be interpreted as a manifestation of belief in another life. The deer cranium and antlers found with the body of a child in the Qafzeh cave and the boar's jaw found in the hands of an adult skeleton at the Skhul rock shelter, both in Israel, have sometimes been considered to be funerary offerings by the proto-Cro-Magnons there. But the same attribution has been made in relation to the mountain goat horns surrounding a Neanderthal boy at Teshik Tash, Uzbekistan; the bear bones carefully arranged with a skeleton in a grave covered with a stone slab in Régourdou, France; the sculpted stone upon the heart of a child in Dedireyah, Syria; the flowers placed on human remains at Shanidar, Iraq; and the hematite powder sprinkled on remains at Le Moustier, France. Of course, one may still seek alternative explanations. So far, no one has presented any definitive proof of ritual or other symbolic behavior before the time of the Cro-Magnons in the Upper Paleolithic. That proof is a much-sought-after prize. It is still up for grabs.

CHAPTER NINE

And the World Was Made Transparent

*And in the mythical societies, said Mircea Eliade, man listened
to the earth because it was not mute, it said things. It was meaningful and intelligible. To understand its language, its structures,
its objects, its life, its rhythms, he resorted to symbols.
Communicating then in the same symbolic code, nature revealed
its mysterious realities: "If the World spoke to him through its
heavenly bodies, its plants and its animals, its rivers and its
rocks, its seasons and its nights, man responded through his
dreams and the power of his imagination... Since the World was
visible to archaic man, he felt that he too was 'watched' and
understood by the World. His prey saw and understood him..., but
so did the rock and the tree and the river. Each had its story to
tell, its advice to give."*

Eduardo Martínez de Pisón, La Protección del Paisaje:
Una Reflexión; (The Protection of the Natural Landscape:
A Reflection)

An Auspicious Geography

All of a sudden and unexpectedly, the spirit of our land, old
Europa, came alive. The rocks, the rivers, the sea, the trees, and
the animals, also the clouds, the sun, the moon, and the stars

above; all sang to humankind, and the wind carried their song. For the first time in their long existence, they finally encountered creatures who understood their message. So they recounted to them their many stories, some sweet, others horrific. But the humans found an ally in nature, a mother who could guide them as they struggled to survive in an often hostile climate. The cycle of seasons and the behavior of animals could finally be explained. Now it was possible to understand and predict natural phenomena.

Millions of years after the first hominids had mastered the art of reading each other's minds, humans learned to read the mind of the natural world, which was then made transparent to them. The eagle is proud and fierce, they said; every creature has its character. The natural stone bridge on the *meseta* became a relic from the legendary days of the giants. Other natural features suggested mythic animals, petrified now, and the human's eternal companions, creatures of legend. Even the great starry fresco of the heavens had tales to tell.

And the people learned to tell those tales and to pass them from generation to generation seated by the ubiquitous fire, to represent them on cave walls and in rock shelters, and to carry them along as stone plaques and statues, or embodied in bone, antler, and ivory, pieces of their animal brothers and sisters. The landscape filled with symbols. For the first time, humans left their mark on nature, and something had changed forever on this planet.

The people felt accompanied, protected, even guided in this world of mythical beings. Life and death finally had meaning. The communion between humans and animals was so intimate that the former knew themselves to be children of the latter and each group adopted a protective totem. We are those humans. We were the first-ever creatures to learn to listen to nature. Old

Shakespeare hit the nail on the head: "We are such stuff as dreams are made of..."

Constructing a Scenario: The Data

Psychologists who study chimpanzees observe a certain parallelism between their learning process and ours up to the age of about two and a half years. After that the gap between us becomes wider and wider until it is a yawning abyss. Chimpanzees continue learning new words up until the age of five, but human children do so at a breathtaking pace, all the while producing ever more grammatical phrases. With these phrases, they give voice to their discoveries and their perceptions of the world they live in and their roles within it. As they learn more and more about other humans and begin to anticipate their actions and reactions, words allow them to see the world as others see it, an extraordinary feat. Children gobble up this new information continually, assimilate it, and become keenly aware of their social environment.

The Neanderthals, on the other hand, were not intellectually like our two and a half year olds. Actually, their physiological development was very similar to ours. To begin with, their maturity at birth was similar to that of a modern child. Like us, they were born much less developed than are chimpanzees. By the age of two and a half they had gone through essentially the same developmental stages leading to adulthood that we have. And they continued growing and learning from adults throughout childhood, just like we do. As adults they acted consciously and purposefully when they manufactured stone tools, built fires, and buried their dead.

The Neanderthals were a species that lived contemporaneously with ours until they disappeared less than 30,000 years ago. They were never a precursor species of ours, never older, and never more archaic. They lived in our time, not in the times of our remote ancestors. But the Neanderthals' forebears in the Middle Pleistocene also acted consciously, as did our forebears of the same epoch, as did the common ancestors of the Neanderthals and modern humans, as did *Homo erectus* and *Homo ergaster*, and even as *Homo habilis* may have done to some extent.

The expansion of cerebral volume in both absolute and relative terms, the differentiation of the two hemispheres, the prominence of Broca's area, the development of the frontal lobe, the preferential use of one side of the body over the other, all of these are phenomena that indirectly indicate cognitive capacities similar to ours. But this is only because the more that two structures resemble each other, the more we are tempted to believe that they have similar functions as well. The function of an organ is really only knowable by analyzing its products. The product of the brain is thought, which is at least expressed and possibly constructed through the use of language.

Unfortunately, it is difficult to tell from the paleontological or archeological record which fossil hominids had language. The only truly direct proof of their capacity to communicate by means of symbols would be the fossil remains of one of those symbols itself. Of course this would not be in the form of a written word, but it would necessarily be a symbol that had been initially conceived and imbued with meaning then agreed upon and codified through extended conversation, a meeting of the minds among people.

Noble and Davidson interpret the population of Australia by members of our species some 40,000, or perhaps as much as 60,000 years ago, as the first evidence of language. This migration required the crossing of a sea, and the construction of rafts or boats for such a voyage. The natural implication is that people must have communicated with each other concerning defined objectives. There is no proof of any previous sea crossing anywhere, although there is some evidence, weak at best, that the Indonesian island of Flores may have been populated 800,000 years ago.

Any kind of long-range economic planning also implies the use of language. We are interested in knowing, therefore, if there were any significant differences between the Neanderthal and Cro-Magnon economies. It has been said that the Neanderthals simply hunted and gathered whatever resources they encountered in their environment, like any animal that lives or tries to survive from day to day. The Cro-Magnons, on the other hand, anticipated the seasonal variations of resource availability and moved freely within large swaths of territory, following the herbivores and seeking the vegetable foods found in different ecosystems. If the mountain will not come to Mohammed, Mohammed must go to the mountain. But this would have required a detailed mental map. Their economy was designed to reap the maximum benefit from what nature had to offer at any given time of the year. In a certain sense, their activities were based on an awareness of natural life cycles, the same principle that governs plant and animal husbandry, i.e. agriculture. This casts the Cro-Magnons as the first biologists.

Finally, there is the question of burial, a symbolic and ritualized behavior if ever there was one, but as we have seen, a topic that stimulates heated debate.

If it is true that consciousness and language are inseparable, then I believe that language has existed since *Homo ergaster*. With the use of computerized tomography, also called CT or CAT scans, we have been able to study in some detail the location of brain lesions in patients suffering from aphasia, the inability to speak. Using the techniques of brain mapping and positron emission tomography, we can localize brain activity during speaking and listening. At the present time, the data obtained from this research seems to indicate that there is no discrete, biological language organ as such. Instead, we find that Broca's and Wernicke's areas communicate extensively with each other, with other regions of the neocortex and with more central and phylogenetically much older brain structures. Some day we may be able to disentangle the nature of relations between the cognitive and language processes and come to a definitive understanding of the extent to which they are or are not one, but that day has not yet arrived.

Up to this point, I have been trying to summarize the evidence available concerning the thorniest problem of human evolution, the development of consciousness, which is the defining characteristic of humankind. I have tried to do so as rigorously as possible. I will use the remainder of this last chapter to relate my particular version of the relevant events based upon the best information now at our disposal.

The Ebro Frontier

Neanderthals and modern humans may have first laid eyes on each other in Israel. This land is just a stone's throw from Africa, to which it is connected by the Sinai Peninsula. Many skeletons have been recovered from two multiple burial sites in Israel that have been repeatedly mentioned in this book. Skhul is a rock shelter on Mount Carmel, and Qafzeh is a cave near Nazareth. The skeletons are about 100,000 years old, the oldest in history, and their anatomy leaves no room for doubt that they are modern humans with just a few archaic characteristics. Some skulls, for example, have a supraorbital torus, the bony ridge above the eye sockets. Because of their essentially modern morphology with just these few archaisms, we call them proto-Cro-Magnons. The skull thickness, the thickness of the bones of the extremities, and the hip structure of the proto-Cro-Magnons are all radically modern and sharply distinguish them from Neanderthals.

A female skeleton and an isolated jaw have been found at Tabun Cave just a few hundred yards from the Skhul site on Mount Carmel. The jaw is about as old as the Skhul and Qafzeh proto-Cro-Magnons, but it is not clear if it belonged to a similar human or to a Neanderthal. Nowadays people tend to think the former. The skeleton, on the other hand, is Neanderthal. Her geological age is uncertain, but she probably lived much more recently. So it seems that the proto-Cro-Magnons arrived in the region from Africa some 100,000 years ago or more and encountered no Neanderthals there.

But Neanderthal remains have been found at the Amud and Kebara Caves, both of them also in Israel and about as old as the Tabun woman, some 60,000 years. Modern humans of that time have not yet been encountered in the region, which makes one think that as the Neanderthals expanded from Europe towards Central Asia and the Middle East, they replaced the proto-Cro-Magnons in Israel or perhaps the latter group had already moved on. In any case, the Neanderthals employed Mousterian industry, as did the Israeli proto-Cro-Magnons, throughout their entire area of distribution. This indicates that the two groups had at least "cultural relations." They both used fire and buried their dead, which could mean that they belonged to the same noosphere, the concept proposed by Pierre Teilhard de Chardin.

The next act of this drama took place in Europe just 32,000 years ago. Modern humans, or Cro-Magnons, occupied almost the entirety of the continent. They had developed a new and varied tool kit that included side scrapers, burins, awls, leaf blades, etc., all produced by retouching long thin flakes of stone that had itself been separated from a markedly prismatic core. Using this lithic tool kit, they were able to manufacture assegai points for hunting from antler, bone, and ivory. As Marcel Otte says, people turned the animals' own weapons, their horns and tusks, against them. These tools, techniques, and products are manifestations of a new technical mode, either invented in Europe or invented elsewhere and brought here. It is Mode IV, the Upper Paleolithic. The first Upper Paleolithic technical complex is known as Aurignacian.

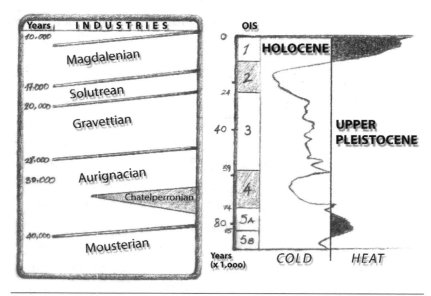

Figure 24: The last 90,000 years. On the left, the technical complexes of the Middle and Upper Paleolithic in the Cantabrian region. On the right, the Paleolithic temperature curves and the oxygen isotope stadials (OIS).

At the very same time, 32,000 years ago, we see spectacular examples of symbolic expression in what is called Paleolithic art, from painted friezes in the cave at Chauvet, France, to animal statuettes of ivory in Vogelherd, Germany, to perhaps the most surprising of all, precisely because of its symbolism, a half-human, half-lion ivory figurine in Hohlenstein-Stadel, Germany. Sculptures at Geissenklösterle, Germany and Trou Magrite, Belgium might even be considerably older.

What was happening to the Neanderthals while all this was going on? A few thousand years earlier, they had had Europe, Central Asia, and the Middle East all to themselves. But by this time, 32,000 years ago, they had lost a lot of territory. The last well-dated Neanderthals were Iberian. It seems that they occupied the entire peninsula except for a band across the far north. The Portuguese archeologist João Zilhão calls the north Iberian geographical boundary between Cro-Magnons and Neanderthals the Ebro Frontier. In general terms, it coincides with the boundary described earlier in this book between two large biogeographical regions, the green Euro-Siberian Iberia and the drier Mediterranean Iberia. According to Zilhão, this is no coincidence. The Cro-Magnons came to Iberia from the northern ecosystems of the Euro-Siberian world, misty forests filled with red deer, roe deer, and boar, and steppes where great herds of horses, reindeer, mammoths, woolly rhinoceri, saiga antelope, and musk oxen grazed. Their forests and grasslands were home to aurochs and bison. Chamois and goats inhabited their rocky heights.

The Cro-Magnons arrived on the European continent at least 40,000 years ago, but they adapted well to the cold and ice, the snow and the fog. The Neanderthals, in the meantime, stuck close by the evergreen forests of holm oak and cork oak, without arctic fauna and perhaps without bison. This equilibrium was upset when the wave of cold blowing over Europe like a frozen wind penetrated deep into the confines of Iberia, drastically affecting the Mediterranean ecosystems and effectively destroying the world of the last Iberian Neanderthals. The people of the steppes swept down in the wake of this climate change and pushed the Neanderthals to the sea.

This scenario is appealing because it links human events to the natural environment. The data available to us also supports the chronology it proposes. On the other hand, it presumes certain ecological developments that we still know too little about. And finally there is an enormous paradox to accommodate. The Neanderthals, a group of humans who had evolved and adapted to the cold on a continent far from the equator, were replaced by other humans recently arrived from Africa.

From the point of view of History with a capital H, we can say that we know what happened. The Neanderthals were replaced by modern humans. There may have been some genetic mixing, but not enough for any Neanderthal genes to reach us. It would thrill me more than anything if I could say that I had even a drop of Neanderthal blood to connect me with those powerful Europeans of long ago. But I am afraid that my relationship with them is strictly sentimental.

In 1999 the Portuguese archaeologist João Zilhão, together with the American paleoanthropologist Erik Trinkaus and other authors, reported the discovery of a child's skeleton in the shelter of Lagar Velho, in Portugal, dating from around 25,000 years ago. This is an individual with generally modern features (ie, of the same type as us) in a context of Gravettian culture (ie, entirely within the upper Paleolithic period). However, the child would show some features which, for these authors, indicate that we would find among its ancestors some Neanderthals—extinct as such three or four thousand years before in the region. This child would be proof that, at least in some places, the Neanderthals and the Cromagnons intersected, so they were not two different species, completely isolated genetically. Thanks to the kindness of João Zilhão, I have had the opportunity to see the

Las Caldas, Asturias
(Magdalenian)
Ivory pendant engraved
on both sides.

1. Torralba y Ambrona (Soria)
2. Áridos (Madrid)
3. Cova Negra (Valencia)
4. Cova Beneito (Alicante)
5. Carihuela (Granada)
6. Zafarraya (Málaga)
7. Atapuerca (Burgos)
8. Los Casares (Guadalajara)
9. Gibraltar.
10. L'Arbreda (Gerona)

11. Abric Romaní (Barcelona)
12. El Castillo (Cantabria)
13. Lezetxiki (Guipúzcoa)
14. Foz Côa (Portugal)
15. Figueira Brava (Portugal)
16. Siega Verde (Salamanca)
17. Domingo García (Segovia)
18. Las Caldas (Asturias)
19. Abauntz (Navarra)
20. Axlor (Vizcaya).

Figure 25: Some of the sites where Paleolithic art discussed in the text has been found. At the top is a Magdalenian ivory plaque engraved on both sides. (Following Corchón, 1997)

remains of that child, although I have not studied them in depth, and of course if I had been the one to find them I should not have dared to go so far.

In addition, I have had the opportunity to study an occipital bone, also from a child and of the same age as that of Lagar Velho, and also found in a Gravettian context, but in the central Mediterranean region of Spain, to be precise, in the Cave of Malladetes (Valencia). The Neanderthals of the area became extinct at more or less the same time as those in Portugal, ie a few thousand years before the child of Malladetes lived. Moreover, the occipital bone of the Neanderthals is very characteristic and clearly different from ours, so that it seems that it would have been possible to recognise in the child of Malladetes some Neanderthal references, if there had been genetic interchanges between Neanderthals and Cromagnons in the Spanish central Mediterranean area. My hope when I approached the Malladetes fossil was that I would find in it some indication of Neanderthal ancestors, but I saw none. The morphology of the child is, in my opinion, entirely modern.

However, this summary accounting of replacement and extinction is only part of the story. The Spanish philosopher Miguel de Unamuno saw capital H History as an inevitable oversimplification and coined the term *intrahistory* to describe the intricacies of human culture and life that form history's "connective tissue." What really happened to people "on the ground"? How and why were historical events produced? To answer these questions, he said, it is necessary to study the multiple stories that comprise history in as much detail as possible. Now, eighty-five years after the appearance of Obermaier's book, this kind of evidence for Iberian History is being deciphered by Spanish and Portuguese scientists.

South of the Ebro, there are no sites with Aurignacian levels more than 30,000 years old. They are all more recent, and they all present characteristics more evolved than the first Aurignacian site in Europe. On the other hand, we do know of a handful of about 30,000-year-old Mousterian sites at Cova Negra, Valencia; Cova Beneito, Alicante; Carihuela, Granada; and Zafarraya, Málaga; and on the Portuguese coast, Figueira Brava, Lapa dos Furos, Pedreira das Salemas, Gruta do Caldeirão. Gruta Nova da Columbeira, also on that coast, may fall into the same category, but there is some doubt about its dating. Several other sites may very well be late Mousterian, such as Cueva Bajondillo, Málaga; the cave at Pêgo do Diablo, Portugal; and Gorham Cave in Gibraltar, at the southernmost tip of the peninsula. It is important to add that several of the Spanish sites have been assigned by pollen or geological evidence to the beginning of the last strong glacial. This confirms that the last Iberian Neanderthals disappeared when dire climatic conditions reached the Mediterranean and Atlantic coasts.

Although the foregoing evidence is more than enough to demonstrate that the end of the Mousterian was late in coming to the temperate lands of the Levant, Andalucía, and Portugal, we still do not know, for lack of reliably dated sites, when the Aurignacians penetrated the high and colder interior of the peninsula. There is an interesting Mousterian site in Burgos, however, called Cueva Millán, dated at about 35,000 to 37,000 years. Another late Mousterian site, Jarama VI, in Guadalajara, indicates an even later end to the Mousterian, about 30,000 years ago. Not only were there no Arctic fauna at Cueva Millán, but steppe rhinoceri were present when the glacial pulsation had already driven them to extinction in other parts of Europe. Given that the *meseta* of the Castilian plateau is in a sense ecologically intermediate between the two Spains, it might be supposed that the replacement of fauna species and the arrival of the Aurignacians would have taken place there before they did near the Mediterranean coast south of the Ebro and on the Atlantic coast south of the Duero. But that remains to be seen.

Although the Aurignacians did not arrive in Mediterranean Iberia until 30,000 or somewhat more years ago, they do seem to have been well established in the northern Cantabrian mountain and coastal band and in Catalonia 10,000 years earlier. The evidence for that is found in sites at l'Arbreda and Reclau Viver in Gerona; Abric Romaní, Barcelona; and El Castillo, Cantabria. Curiously, there are no earlier dates for the Aurignacians in Europe other than some questionable finds in Bulgaria. All the data seem to indicate that the Cro-Magnon colonization was very rapid, taking place some 40,000 years ago, but that the Neanderthals were not immediately eliminated. Instead there was a long period of coexistence, during which the Cro-Magnon and Neanderthal populations were in irregular contact with each other.

There are two ways to imagine this coexistence. At lectures I illustrate them with my two hands raised and open, one for the Cro-Magnons and the other for the Neanderthals. One possible model is illustrated with the fingertips of one hand touching those of the other, the Cro-Magnons to the north and the Neanderthals to the south. This would have been the case on the Ebro Frontier and perhaps in Italy and the Balkans, these being the other Mediterranean peninsulas, and on the Crimean Peninsula in the Black Sea. I illustrate the alternative model by interlacing my fingers to indicate what may have been the situation in the non-Mediterranean, Euro-Siberian part of the continent, where Neanderthal and Cro-Magnon populations were interspersed for thousands of years. Cova dels Ermitons, Gerona is an archeological site in the green northern region of Spain that suggests that Mousterian Neanderthals may have continued to live there in isolation for several thousand years after the first

Aurignacian Cro-Magnons arrived in Catalonia. There are a number of digs in France that exemplify an Upper Paleolithic technology called Chatelperronian. Its practitioners also elaborated their tools out of long, thin flakes of stone, also used bone to produce assegai points and needles, and also made personal adornments out of ivory. We use the term Chatelperronian to describe Spanish finds as well. On this side of the Pyrenees, they are found at sites at Ekain and Labeko Koba, Guipúzcoa, and at Cueva Morín and El Pendo, Santander. The most exciting thing about the Chatelperronian finds is that human remains associated with them have been recovered from a couple of the French digs. From Saint Cesaire, we have a Neanderthal jaw and a large part of its cranium. This individual was absolutely "classical," with no modern or even intermediate features. Very fragmentary but identifiably Neanderthal remains have been found at Grotte du Renne, in Arcy-sur-Cure. Perforated and grooved teeth and bones already prepared to be hung or strung together were found in association with Chatelperronian tools at the latter cave, as have ivory beads and rings and fossils of sea creatures, all of which these people used for personal ornamentation. In fact, the name of this book was inspired by this surprising discovery. The Neanderthals at Grotte du Renne wore necklaces. At Quinçay, another French Chatelperronian site, researchers found six teeth perforated at the root. Other Mode IV, or Upper Paleolithic, industries practiced by Neanderthals in many parts of Europe were somewhat parallel to the Chatelperronian. Just to provide three examples, in Italy the corresponding industry is called Uluzzian; in Central Europe it is called Szeletian; and in Bulgaria it is called Bachokirian.

There is currently a lot of interest in ascertaining more accurate dates for these earliest Upper Paleolithic finds. Taken together, they run from 30,000 to 40,000 years ago. Many authors consider all the Chatelperronian sites to be subsequent to the first Aurignacian ones, and believe that the Neanderthals learned tool manufacturing and self-adornment from the Cro-Magnons. In three sites, the rock shelters of Le Piage and Roc-de-Combe in southwestern France and the cave at Pendo, it has been thought that the Chatelperronian level was sandwiched between two Aurignacian levels, one below it and another above it, as if after the Cro-Magnons initially arrived, they were replaced by Neanderthals for a period, but later returned to stay. Respectable arguments have been made to challenge this interpretation, however, so there is some doubt about the presumed Chatelperronian-Aurignacian interstratification. In fact, some authors believe that the development of Chatelperronian industry preceded that of Aurignacian, and that the Neanderthals were the true inventors of Mode IV, the Upper Paleolithic, while the recently arrived Cro-Magnons were mere imitators. According to this argument, that is why there is no previous sign of Mode IV technology outside of Europe to demonstrate that the Cro-Magnons arrived having already developed it elsewhere. Finally, it is worth asking if the Cro-Magnons and Neanderthals could not have developed the new technology in parallel or under each other's influence.

In order to demonstrate that the Cro-Magnons brought the new technology to Europe with them, we would want to find previous evidence of Upper Paleolithic innovations elsewhere. The best place to look would be in Africa, whence, it is thought, modern humans migrated. Little such African data is available, but we do have some indications, less than definitive, of what could turn up. From the Enkapune Ya Muto rock shelter in Kenya, we have some 40,000-year-old beads made from the shells of ostrich eggs. From Blombos Cave in South Africa, we have evidence of bone technology some 80,000 to 95,000 years ago, and from

Katanda in the Democratic Republic of Congo, we have barbed harpoon points of the same age.

If we are to question the available evidence, we may also want to ask whether the Cro-Magnons were necessarily the originators of the first Aurignacian industry in the north of Spain. Could it not have been the Neanderthals? Federico Bernáldez de Quirós and Victoria Cabrera, who work at the classic site of El Castillo, first excavated by Obermaier in 1910–1915, see no difference between the economy, the way of life that is, of the Mousterian Neanderthal occupants of the cave and that of the people at Aurignacian levels immediately above, dated at about 40,000 years of age. They also report significant continuity in the lithic tool kits. Could these have been the same people? At l'Arbreda, in contrast, Mousterian products were almost all made of quartzite, while the Aurignacian items were of imported silex.

Actually, we do not yet have enough human remains from that first Aurignacian period to answer all our questions. The oldest Aurignacian human remains are those from Mladec, in Moravia, a region of the Czech Republic. They are anatomically modern and probably about 32,000 years old, about the same age as the first well-dated manifestations of figurative art. That was a time when Neanderthals still lived south of the Ebro and perhaps in other parts of Mediterranean Europe as well. Human fossil remains from Brno, also in Moravia, may be approximately the same age, but they were not found in association with any industry. A modern frontal bone at Hahnöfersand, Germany, although lacking contextual associations, has been closely dated to just a few thousand years earlier. The Cro-Magnon fossils themselves might be contemporaneous with those of Mladec or perhaps somewhat later, which would make them late Aurignacian, at about 30,000 years. There are many subsequent European fossil remains after that time, all of them of modern humans, and all associated with post-Aurignacian technical complexes known as Gravettian, Solutrean, and Magdalenian. We should take note of the ecological difference between the French Chatelperronian Neanderthals and the Aurignacian moderns in Moravia on the one hand, and the contemporaneous Mediterranean Neanderthals on the other. The first two peoples lived with reindeer, mammoths, woolly rhinoceri, and cold-weather fauna unknown to the third.

In any case, it seems that typical arctic mammals had become uncommon over the entire peninsula north of the Ebro in the period between 30,000 and 40,000 years ago. They have only been reported in Guipúzcoa, which due to its geographic position virtually mirrored conditions in Aquitane during glaciations. The Neanderthals were alone throughout the glacial maximum prior to the arrival of the Cro-Magnons. That was OIS 4 on the marine isotope scale. The arrival in Europe of the Cro-Magnons and their at least 10,000-year-long period of coexistence with the Neanderthals took place at the end of an interstadial, OIS 3, a relatively warm and humid period between the two coldest and driest stadials, or pulsations, of the last glaciation. With the arrival of OIS 2, the subsequent and even more bitterly cold maximum, the last Mediterranean Neanderthals disappeared. We have to keep in mind that an interstadial is never as mild as an interglacial, a warm period between glaciations. We are currently experiencing an interglacial, OIS 1.

This is the current state of our knowledge, which I and most authors choose to interpret in the following terms. Modern humans were the innovators of Aurignacian industry from its beginnings. Neanderthals in certain areas learned from the Cro-Magnons and reproduced, all in their own fashion, the new stone manufacturing techniques, the use of animal-based materials, and the love for personal adornment. Other Neanderthals south of the Ebro did not adopt these changes, maintaining their culture right up to the time of their extinction, which coincided with a glacial advance and the ecological change that it brought about.

As can be appreciated, we need to fill in the gaps of the relevant data and learn more about the regional details that will help us to understand in a convincing fashion what it was that sealed the Neanderthals' fate.

The Color of Their Heather

The usual media for prehistoric artistic representations were stone walls, stone plaques, and pieces of bone, antler, or ivory. Certainly wood was also used, although of course it perished over time. But the Cro-Magnons used another very special "canvas," their own bodies. I'm sure that they painted themselves, although there can be no direct evidence of the practice. In some burial sites, they seem to have spread blood-colored ochre powder quite liberally. Ochre has excellent antibacterial qualities, so it would have preserved the leather and skins that prehistoric people wore. We cannot be completely sure if they were thinking more of their clothing or of their bodies or even what their goal really was. It is also possible, but not definite, that Neanderthals also used red ochre in their burials, as in the case of Le Moustier, or to paint themselves while living. Blocks of hematite have often been found at Neanderthal sites. At level 10 of the Grotte du Renne alone, the level that contains the most objects of personal adornment (twenty-four), the Neanderthals had carried nearly forty-five pounds of red ochre into the cave.

What is doubtlessly new and characteristic of the Cro-Magnons is the sheer quantity of personal ornamentation that they used. They hung these objects from their necks. They were strung as necklaces, belts, armbands, and bracelets, and they were sewn onto skins worn as clothing and onto hats. These decorative objects were so varied that it is not always possible to distinguish them clearly from pieces of what is called portable art. We could establish a theoretical difference by saying that personal adornments belonged to individuals and portable art belonged to the collectivity, but many of the famous Paleolithic Venuses, statuettes of abundantly fleshy women that are characteristic of the Gravettian period, were probably worn around Cro-Magnon necks as well. Humans are known to have possessed these Venuses all the way from the Pyrenees to Siberia between 20,000 and 28,000 years ago. Curiously, they were absent from the peninsula. The Venuses themselves are portrayed wearing sculpted adornments as well, including necklaces, bracelets, head ornaments, and more. People of the Magdalenian period profusely decorated utilitarian objects also, particularly their spear throwers and ceremonial staffs, which also had holes into which assegai points were inserted for straightening. The points themselves were manufactured from pieces of deer antler or of bone. Any natural curvature would be corrected as part of the production process. Modern populations practice the same technology. As Kaj Birket-Smith, the expert on the Inuit, tells us, "To writers on prehistory, it is of considerable interest to know that bone points are made by first softening the material in hot water, then straightening it and giving it the desired shape with a tool made from a piece of caribou horn with one or more holes in it." The most elaborately decorated spear throwers and ceremonial staffs were most assuredly objects of prestige above all else, with symbolic meaning either to the whole group or exclusively to individuals of elevated social position. In the latter case, the staff would contribute to the social status of one person in particular.

Objects of adornment were frequently made out of animal parts, such as fox or deer canines, bovine or cervid incisors, or mollusk shells, which are sometimes found at sites far from the sea. Burials in the Italian province of Liguria were accompanied by large numbers of shells. Eight shells have been recovered from the Aurignacian level of the l'Arbreda dig, and one of them has two perforations for hanging. Unlike the image made popular in movies, people did not primarily adorn themselves with the large canines of fierce animals like bears, lions, or leopards, or even wolves. North of the Pyrenees, the more modest canines of polar fox, even smaller than those of the common fox, were the most popular. They were perforated at the root for stringing. The Chatelperronian Neanderthals of the Grotte du Renne and Quinçay used them also. It seems that they saw something special in the arctic fox that escapes us today. It is a variable species, white in the winter and grey-brown in the summer, but never a very imposing sight. It is not useful as food, but people may have used its pelt.

At times people put a lot more effort into the adornment of their bodies. Beads are a good example. Being so small and so numerous, they required the most time to produce. They were made of bone, ivory, antler, or soft stone. Many kinds of carved and sometimes elaborately decorated pendants were made of the same materials. A triple burial in Sungir, Russia takes the honors in this department, hands down. A sixty-year-old adult and two adolescents, a boy and a girl, are buried with an astounding number of adornments, representing thousands of hours of work. The production of ornaments there on the steppes was much more laborious than in Liguria, where all the shells people could want were available for the taking, and all they needed to do was string them. The adult at Sungir had 3,000 beads of mammoth ivory sewn onto his clothes and hat of skin, the boy had 5,000, and the girl had even more. The boy also wore a belt decorated with 250 arctic fox canines, and the grave contained many other items such as bracelets, pendants, assegais, ceremonial staffs, etc., really an interminable list.

The functional motivation for adornment was not purely aesthetic. It visually communicated very important information about the individuals who displayed it, just as today the way that people cover their bodies expresses their affiliations and status, including group membership, social position, marital status, and condition as widowed or widower. The human brain has a limited capacity to remember and identify people. Although it is said that Napoleon knew every one of his soldiers personally, that did not keep them from wearing identifying clothing, their uniforms that is, or motivate them to remove their unit insignia, symbols of rank, individual medals, or other signs of distinction. Social identity and condition, as well as hierarchical position within a group, are expressed by means of visual symbolism, by appearance. Despite being the only species that has the use of oral language for communication, we are also, surprisingly, the species that communicates the most visual information by means of personal appearance. Paradoxically, we members of the talking species do not need to open our mouths in order to know who is in front of us. But the paradox is only apparent. The symbols of the visual code are really another form of language agreed upon and communicated through the use of words. They are understood as such when they are conceptually decoded.

The visual code is not intended to associate names and faces so much as to facilitate social relations, to establish interpersonal connections of camaraderie, and to form cohesive groups with common objectives. In this we really are limited by biology to groups of no more than about 150 people, perhaps a few more or less, as I discussed earlier. Although religions preach universal love, we are not really able to love strangers with the same intensity that we love our friends and neighbors. Personal appearance, what is called our *image*, infinitely increases the size of our social groups, which then include many people whom we do not know personally but whom we can somehow know on the basis of their appearance. The individual transfers his or her identity to objects of adornment, which then *em-body* that identity. They become a part of that person and the social meaning of his or her body. At the same time, the adornments become a medium through which corporal expression becomes more articulate. As Yvette Taborin says, ornamentation amplifies the body.

Over time, each group determines its signs of identity as expressed by objects with symbolic value. Today there are different ways of dressing depending on one's political ideology or group, i.e. *tribe*, that one wishes to identify with and that one prefers others to identify him or her with. But the symbolic code can only be read by members of the same society, because symbols are, by definition, arbitrary conventions. Like words, they are tacitly agreed upon and shared by the collectivity. Until very recently, it was socially obligatory for widows and widowers in Spain to indicate their condition by wearing black. But it was not always thus. Isabel the Catholic decreed that black clothes be worn to express mourning simply to avoid the economically wasteful use of expensive white fabrics that had been fashionable up until that time. The different Scottish clans distinguish themselves by the color of the heather that they wear on their caps, not by the tartan, as many believe. For example, red heather and white heather may indicate different clans. And I hardly need mention the use of color combinations to express national sentiment or an identification with sports teams, sometimes both.

Ethnicity

There has been an explosion of symbols of every imaginable kind. They are everywhere; they completely surround us. But our species is now characterized not only by the mere existence of symbols for communication, but by their use as powerful media of social integration. Any symbol is by definition the patrimony of the community that has created it and that understands it. Since the group communicates consciousness to the individual by means of language, we can say without too much exaggeration that there is also a supraindividual consciousness. In a sense, natural selection works at the level of the individual. After all, individuals compete among themselves on the basis of their individual characteristics. The lucky ones survive longer and reproduce more, passing their characteristics along to posterity through their descendants. But since human evolution is also a history of competition and selection among groups, only the most successful groups win out and perpetuate themselves. At the end of the day, it is not the individual, but the group that really counts most. To whatever extent Cro-Magnons lived in solidarity with other members of their group, they were equally ruthless towards outsiders.

As a rule, the Neanderthals did not use personal ornamentation. In fact, it has been found only at the two Chatelperronian Neanderthal sites of Grotte du Renne and Quinçay. Perforated shells have also been found at a couple of Uluzzian sites. Uluzzian industry was a sort of Italian equivalent of Chatelperronian, also presumably practiced by Neanderthals. No ornaments have ever been recovered from any of the Mousterian levels though, neither those from the time before the arrival of the Aurignacians nor those from after it. The Neanderthals south of the Ebro, for example, disappeared without ever having taken up the practice of self-adornment. The discussion above on the origin of Chatelperronian industry applies equally to its associated ornaments. Francesco d'Errico, João Zilhão, and other archeologists believe that the ornaments at Grotte du Renne are as old or older than any found at Aurignacian levels, but many other authors, including Randall White and Yvette Taborin, recognized specialists in prehistoric ornamentation, believe that the personal ornaments at Kostenki 17 in Russia and Bacho Kiro, Bulgaria are both thousands of years older. The most widely accepted hypothesis today is that the Cro-Magnons invented personal ornamentation and some of the Neanderthals imitated them, just as they imitated their stone working techniques.

I concur with this hypothesis, but two diametrically different conclusions may be drawn from it. One is that ornamentation was not meaningful to the Neanderthals. They did not have the capacity to capture the symbolism behind it for the simple reason that it amounted to a visual language, and the Neanderthals, like modern chimpanzees, did not have the capacity for symbolic language, whether oral or visual. In this view, their brains had developed to favor "natural" intelligence, a kind of "instinctive intuition," but had not attained the capacity for abstraction or symbolic communication, in this case through the use of objects. They imitated the Cro-Magnon practice of ornamentation, but without understanding it. The other possible conclusion is that the Neanderthals did have a fully modern capacity for language and for the use of symbolic communication through objects and otherwise, but they became extinct before having the opportunity to develop them to the extent that we did.

I believe that the truth lies somewhere between these two hypotheses. I am not trying to conciliate the parties. In fact, I can say that both of them are partly wrong. The Neanderthals had the technical capacity to manufacture tools out of stone and bone in the same fashion as the Cro-Magnons. The proof of that is that they did so. I believe that they also had language and practiced funerary rituals. They were humans, not just in the taxonomic sense of belonging to the same evolutionary group and sharing many genes with us, but in the more spiritual sense of belief and emotion, aspects of the human mind. Our human condition did not just fall from the sky. It was made possible by abundant precedent, because many steps had been taken in our direction. Nonetheless, the Neanderthals did not develop our extreme specialization in the production and employment of symbols and they did not achieve our boundless level of creativity. They were not so carried away by their own imagination. They were deep-seated realists, if you will, which did not make them inferior.

Some, of course, believe that the fully human condition, including the use of language and associated forms of symbolic expression such as art, was born with our species 150,000 to 200,000 years ago. But those who take this view still wonder why modern humans waited so long to leave Africa and eliminate competing humans (not exactly humans, according to them). And then, they ask, after the moderns put in an appearance in Palestine 100,000 years ago, why did they retreat from the Middle East, abandoning it to the Neanderthals? Some say that although the proto-Cro-Magnons were anatomically, or at least skeletally, modern, a few of their neuronal circuits had yet to be completely connected before they could become fully conscious and fully Cro-Magnon.

I do not see it that way. As I have said, I believe that our diminished robustness was closely related to the appearance of the capacity for articulated language, and that modern humans were completely modern from the beginning. My answer to the question of why they did not eliminate the Neanderthals in one fell swoop is that the Neanderthals were also human and very intelligent at that. In fact, modern humans may have arrived in Australia 60,000 years ago, long before they set foot in Europe. They may have encountered the last of *Homo erectus* on the way across Asia, possibly the population at Ngandong, Indonesia. It is also possible that *Homo erectus* was unable to put up a resistance as effective as the Neanderthals'. Our hypertrophic capacity for symbolic communication and linguistic articulation is very useful for telling stories, but that does not mean that it would necessarily have given us a decisive advantage over another human group, the Neanderthals, who were very strong and better adapted to European climates and natural environments.

The competition between the Neanderthals and the Cro-Magnons may have lasted for thousands of years thanks to these Neanderthal advantages, and our ancestors may have triumphed due to two other factors. In the first place, they invented and consistently improved the new Aurignacian technology for the production of tools. They already had this technology when they entered Europe, which afforded them a certain superiority that they had not enjoyed in earlier encounters with the Neanderthals. Some Neanderthals were able to take up the new technology, those who lived in higher density populations and those who were able to hang on for a period of time despite being surrounded by Cro-Magnons. Nevertheless, the Aurignacian technology gave the Cro-Magnons an initial advantage that enabled them to reach the north of Spain very quickly.

Paradoxically, the second factor was the climate. Although the most bitter glacial conditions had not yet descended upon Europe 40,000 years ago, the center and north of the continent were cold, and enormous herds of reindeer and mammoth roamed the vast steppes. The Neanderthals were biologically better adapted to the cold, but the Cro-Magnons' symbolic systems permitted them to make very effective cultural-ecological adaptations and form alliances among groups separated by great distances. This was accomplished through the sharing of old myths that linked them with the natural world and with their common ancestors. Telling stories was their specialty, and their old myths were oft-told collections of stories. As environmental conditions continued to deteriorate, reducing the human population in numbers and in density, this advantage took on added importance. Finally, even the stable Mediterranean world was dramatically affected by climate change. Forests gave way to steppes and large herds of horses, soon to be followed by horse hunters, none other than our ancestors the storytellers.

A simple way to find out how the two human groups confronted the hostile climate is to travel to the East European Plain, stretching from the Carpathians in the west to the Urals in the east and to the Arctic Sea in the north. In the south, the vast low-lying plain ends at the Black Sea, the Caspian Sea, and the solid wall of the Caucasus that closes off the space between them. There are no significant elevations in the entire area, so the severity of the climate intensifies as one travels northward. In this warm period of ours, the average January temperature at the center of this area, also known as the Great Russian Plain, is fourteen degrees Fahrenheit. That is at fifty degrees north latitude. It is by no means a hospitable place to spend a night in the open.

The first humans who dared to migrate into the Plain were Neanderthals. That was just about 120,000 years ago, in the interglacial period that preceded the last glaciation. They reached a point above fifty degrees north latitude, according to the evidence at the Rikhta, Zhitomir, and Khotylevo I sites. In fact, Khotylevo is at fifty-two degrees north! There is no doubt, then, that they were able to adapt to extreme conditions, and it is difficult to deny them an extraordinary aptitude for organization and planning. Would human existence in such circumstances be possible without it?

However, the coming of the last glaciation forced a Neanderthal retreat to the south, where they took refuge on the Crimean peninsula and the northern slopes of the Caucasus. The last of them probably disappeared there at the same time as their Iberian homologues, between 25,000 and 30,000 years ago. The Cro-Magnons, in contrast, succeeded in conquering the Great Eastern Plain. They arrived at the site called Kostenki 17, at a latitude of fifty degrees north, between 35,000 and 40,000 years ago. Their success, where the Neanderthals were defeated by the cold, can be partially attributed to their superior technology. At Kostenki 14 there is a large number of bone awls and needles that these people used 30,000 years ago to improve the insulating efficiency of the skins they used to protect their bodies from the elements. The tailors of the day may well have produced garments as warm as those of the modern Inuit.

Later, when the last ferocious glacial maximum drew near, the modern humans of the Great Eastern Plain learned to build shelters framed with mammoth bones and covered with skins, to keep their hearths burning consistently within them, and, in the absence of other fuels on the inhospitable plain, to use mammoth bones for that purpose as well. By the time that the cold reached its fiercest point beginning 25,000 years ago, modern humans were prepared to survive the worst. The average January temperature 20,000 years ago must have been unbelievably low, and the desolation of the northern world must have been terrifying.

But besides their technology, which consisted of tools, clothing, and effectively heated shelters, other seemingly more humble objects found at Kostenki 17 doubtless contributed to human survival on the Great Eastern Plain. I am referring again to their personal ornaments. They tell us that their proud users had created the new social dimension of *ethnicity* that would forever mark human destiny. An ethnicity is a group that is differentiated and identifiable by more than just the purely biological. It is organized around shared symbols.

Once liberated from competition with other species, *Homo sapiens* was fruitful and multiplied, developing ever more efficient and deadly generations of technology. While the Lower Paleolithic Oldowan and Acheulean stone technologies, and even the Mousterian and other Middle Paleolithic technologies, had been uniform over immense geographical areas, those of the Upper Paleolithic were more diverse in both design and in regional variation.

Jean-Pierre Bocquet-Appel has a demographic explanation for this phenomenon. Since the Neanderthals and other "archaic" human species never attained high population density, neighboring groups were obliged to exchange individuals in order to avoid extinction, thus spreading limited biological and cultural resources very thinly. But beginning with the demographic explosion of the Upper Paleolithic, modern humans formed ever-larger groups. As these large and dense population clusters became reproductively viable and economically self-sufficient, they also became more biologically and culturally closed.

The Hellenes of the Second Persian War saw themselves as members of one community because they shared an ancestry and a language, they performed the same rituals, they regarded the same places as holy, and they had similar everyday habits. Symbols and more symbols.

The shared stories and myths that were so useful to small and dispersed bands of Cro-Magnons grew to be insurmountable barriers when their populations grew and turned their backs on each other.

The final result of our evolutionary history is that two identities coexist within each one of us, one individual and the other collective. To deny the existence of either of these two aspects of human nature is to close one's eyes to reality. While our individual identities promote egoism and neutralize the impulse to social solidarity, the collective identity can lead to the abyss, because it makes us easily manipulated. In the recently ended century alone, the bloodiest in human history, tens of millions of people died in confrontations between symbolically defined group identities. At the same time, any deviation from intragroup unanimity or resistance to necessary social homogeneity has been viciously persecuted as an intolerable threat to the collectivity. Have we reached an evolutionary dead end, or is it possible that humans may someday overcome this abiding contradiction between the individual and the group? The answer, my dear reader, is blowin' in the wind.

Domesticated Man

And the days passed. And the years.

And Death came and swept them from their refuge; all of that race disappeared with all of its tales and all of its history.

But all things came back to life in that place. Other trees stood tall and other men bent to the ground. Newborn litters roiled in the caves; the tapestry never unraveled.

Wenceslau Fernández Flores,
El Bosque Animado (The Animated Forest)

In the beginning, between five and six million years ago, there was the ape. Or rather, the ancestor that we share with the chimpanzees, a denizen of the African rain forest. This animal was on the verge of consciousness, above all in its social existence. The hominids and the ancestors of the chimpanzees later appeared in various parts of tropical Africa.

As climates and ecosystems changed, the hominids adapted to life in ever drier forests, while the chimpanzees' ancestors remained in the rain forest. Some hominids were already bipedal over four million years ago, though they were still very much forest dwellers. They were almost completely vegetarian, "almost" since chimpanzees do consume insects and hunt small mammals when they can.

Two and a half million years ago, evolution produced a hominid species, *Homo habilis*, with a larger brain. *Homo habilis* hit one rock with another to produce a sharp edge. This was important because the function of this first-ever blade was to cut meat. A significant change of diet was made possible, and with it a new ecological niche.

A short while later, in geological terms, a really new hominid appeared, *Homo ergaster*. This species' brain was much larger than any chimpanzee's, and it matured more slowly. *Homo ergaster* manufactured tools according to a standard design, and individuals communicated with each other by means of symbols. At the very least, they could control their emotional expressions, both gestural and oral. Their expressions were no longer mere symptoms or indications of mood. They were signs used by individuals to transmit the information that they *wanted* to transmit, only when and to whom they chose to do so, an emergent language. In addition, their young had a learning period long enough to develop their cognitive capacities far beyond those of the chimpanzees. If we could administer the intelligence tests we use on chimpanzees to *Homo ergaster*, they would get much higher scores than the chimps.

These hominids of the species *Homo ergaster*, or let us say these humans, began to create a social and cultural environment around themselves that afforded them ever more independence from the physical environment. Thanks to this development, their populations grew, and they were able to spread across Eurasia probably more than 1.5 million years ago, successfully overcoming the difficult climatic and ecological challenges of latitudes far removed from the equator. Their hominid contemporaries, the paranthropoids, never accomplished that. In fact, they never left Africa.

Things went so well for these humans outside of Africa that they populated almost all of Asia and Europe, reaching the cold lands of Germany and England half a million years ago. Others had long before reached the Iberian Peninsula in the far west and China and Java in the far east. Their oldest remains have been found on Java. They are called *Homo erectus*, although they did not greatly differ from the African *Homo ergaster*.

The humans isolated in Europe evolved to produce an indigenous species, the Neanderthals. The Neanderthals retained great physical strength and were physiologically well adapted to the European climate. They had large brains, which they used to communicate among themselves, to make controlled use of fire, and to produce elaborate tools whose manufacture required many steps. They also used their ingenuity to resolve the difficulties of survival in the markedly seasonal European ecosystems, hardly suitable in themselves for hominid life.

While the Neanderthals evolved in Europe, we did so in Africa. But 300,000 years ago, at the time of the Sima de los Huesos, our ancestors and those of the Neanderthals were neither physically nor behaviorally very different from each other, because they had not evolved separately for very long. Then the second great expansion of the human brain took place, and since it occurred independently in Europe and Africa, it produced somewhat different results.

The results that we are most familiar with are those of our own lineage, since they are staring us in the face and in the mirror. One of them is our fabulous articulated language, at the service of a unique capacity to manipulate symbols, or to put it another way, to tell stories and create fictitious worlds. Creativity is our hyperspecialization. It arose only in the African branch of human evolution, not in the European one. The Neanderthals advanced beyond their ancestors at Sima de los Huesos within an unchanged cognitive and communicative paradigm, already very advanced indeed. But they did not develop, as we did, a revolutionary new system to communicate information.

In order to demonstrate to yourself the extreme specialization implicit in our use of language, I propose that you try an experiment designed by Philip Lieberman. Read any text out loud for ten seconds. You will see that you can easily read an amount of text composed of about 200 letters, or twenty per second. Although these letters do not represent exactly 200 sounds, or phonemes, this should give you an idea of the human articulatory and acoustic capacity. It is amazing that we can produce and distinguish sounds at such a pace.

The mind of modern humans was also different from that of their Neanderthal contemporaries, but not because they were more intelligent, at least not technically so. Their way of seeing the world was thoroughly transformed. This was due to a genuine aberration, described astutely as always by Konrad Lorenz.

All animals filter the stimuli that they receive. We receive so much information by means of our senses that we need to rapidly, automatically, and unconsciously select only the most important data from among it. Only the stimuli that pass through a mental filter produce reactive behavior. If we did not have a filtering mechanism for this purpose, we would not be able to do anything, because we would spend our whole lives analyzing the data that we receive.

We modern humans are extraordinarily social primates, and even extraordinarily social humans. We are very attentive to the signals produced by other humans, which help us to read their minds and predict their actions. For the sake of efficiency, we react quickly to simple and isolated stimuli. We scrutinize the faces of other humans as though all social existence were a game of poker, and we do so with such intensity that we can detect the slightest indication of change.

And that is the key to why the natural world took on a spirit, or took on spirits, of its own. We have an enormous capacity to analyze reality into ever-smaller factors. This capacity is so acute, in fact, that despite our superb cognitive faculties we fall into resounding errors of interpretation that no other animal could possibly make. We attribute emotional value and human characteristics to the most surprising objects.

As Lorenz said, "Steeply rising, somewhat overhanging cliff faces or dark storm-clouds piling up have the same, immediate display value as a human being who is standing at full height and leaning slightly forward, demonstrably menacing." The bony ridge protruding above the eagle's eyes seems to be a furrowed brow, and together with the pulled-back corners of the mouth, it gives the bird a look of stubborn determination. Since the camel and llama hold their heads high with their nostrils above the level of their eyes and the corners of their mouths down low, it seems that they look at us with disdain, so they strike us as unpleasant creatures.

We also attribute aesthetic qualities to animals. The hippopotamus is ungainly; the flamingo graceful and elegant. What is even more significant is that we assign them moral values. In children's stories, the wolf is the bad guy and lambs are always good. The ant is industrious and the cricket is lazy, etc. To sum this up through a common childhood experience, Lorenz relates that as a boy, he felt that a streetcar with half-drawn window shades was watching him menacingly. The eyes play such an important role in our interpretation of faces that any object with orifices, as in the case of the streetcar or a house with windows, tends to remind us of a face. We then assign these features to a category, like "friendly" or "unpleasant," based on the disposition of the elements that surround the supposed eyes, which we immediately interpret as nose, mouth, eyebrows, forehead, and hair.

This curious and erroneous perception of animate and inanimate entities as human and the capacity to tell stories in which those entities appear are what brought nature to life for us. The great advantage that we humans have derived from this peculiar defect is that it helped us to understand natural phenomena. If individual consciousness emerged because it is useful to see things from others' points of view, assuming that they too are conscious, then putting ourselves in the place of other natural entities and attributing consciousness to them as well is an unscientific but effective way to do biology and geology. Geography too. The best way to remember a map and to share it with other members of the community is to associate topographical features with people and stories. From La Granja de San Idelfonso, where I have spent many of my summers, one can see an enormous mountain called *La Mujer Muerta* (The Dead Woman) named for its shape. Another mountain nearby is known as *El Montón de Trigo* (Wheatstack).

At the same time that this marvelous aptitude was developing, another notable change, at first glance unrelated, was taking place. Our skeletons were becoming more gracile. Our hips narrowed, saving us energy with every step. The Neanderthals and the modern humans were different physical types, but that is not the whole story; there were cranial differences, and we had different capacities to articulate a range of sounds. I do not agree with Dr. Ian Tattersall, of the American Museum of Natural History, that the large brain and articulate language were exaptations, features that emerged independently of each other and were unrelated to the manipulation of symbols. I believe instead that they were true adaptations, mutually dependent adaptations in fact, because they need each other to function.

Since *Homo habilis*, we humans have specialized in intelligence, just as birds have specialized in flight. We had already come a long way in that direction when we colonized Europe for the first time. Although that European population became isolated, the demands of survival and competition among groups stimulated the continued development of intelligence without compromising physical strength. The Neanderthals had bodies as powerful as those of their ancestors and a larger brain to boot.

Our African ancestors were also strong and grew ever more intelligent. At some point though, in some human population, a physically weaker but more highly communicative variant appeared. It may seem surprising, but the two characteristics are related. Our improved articulatory phonetics were made possible thanks to a reduced face size, and this was only possible thanks to reduced respiratory demand. The Neanderthals were really massive, with an enormous thoracic and pulmonary capacity. The large volume of air needed to oxygenate their muscles needed to be warmed and moistened in their oral and nasal cavities before entering their lungs, so the horizontal segment of the vocal tract maintained its length. The cold European climate that they lived in contributed to this need.

The first modern humans in Africa were surrounded by other populations as robust as the Neanderthals, but they took a different evolutionary route, an alternative strategy to solve the same ecological problems. They developed a brain specialized in the manipulation of symbols, a less protuberant face that perhaps entailed a greater risk of choking but provided a magnificent articulatory instrument, and a body less powerful when sudden concentrated force was called for but more energy efficient in the long-run and therefore better suited to extended migration. These changes took place some 150,000 to 200,000 years ago, and according to molecular biologists, affected only a small portion of the African population. Their finding of minimal genetic variation among current human populations leads them to that conclusion. Despite our differences of skin color, hair texture, the shape of our eyes, and a few other characteristics, we are all very similar. Racist theories are not only a moral abomination, but they are without scientific merit. That small African population from whence we sprang may have included 10,000 or 15,000 members, equal in number to the Iberian population at the time.

Neanderthals and modern humans were alternative human models, each representing a different but effective evolutionary response to the identical challenges they faced. Both they and we grew in number and expanded our territories. The Neanderthals moved beyond their European homeland, and we left ours in Africa. It was only a matter of time until we would meet.

Between 1929 and 1931, Lluís Pericot found some 5,000 painted and engraved stone plaquettes in a cave at Parpalló, Valencia. They had been produced by Paleolithic humans over the course of thousands of years. The plaquettes have recently been studied in depth by Valentín Villaverde. The animals depicted, in order of frequency, are goats, horses, deer, and aurochs. There are also four chamois, four boar, three wolves, three foxes, a lynx, a mustelid (possibly an otter), a partridge, and a duck. There is not one representation of cold weather fauna, be it mammoth, woolly rhinoceros, or reindeer, and none of bison, which perhaps never established themselves in the eastern part of the peninsula.

There is another category of representational art found in the eastern peninsula and consequently known as Levantine art. It is concentrated in the area around Valencia, but has also been found in Andalucía, Aragón, Castilla-La Mancha, and Catalonia. It is found on open air rock faces and in rock shelters, in the daylight rather than in the darkness of caves. Levantine art depicts animals and people, both men and women. The people are often engaged in hunting, gathering, or possibly ritualistic dancing. Since the discovery of these representations, their age has been sharply debated. Obermaier and Hernández-Pacheco, by the way, participated from opposing camps. Today it seems clear that they are not Pleistocene, as Obermaier believed, but postglacial. Many of the paintings illustrate the hunt. Participants are armed with bows and arrows. The prey are deer, bovines, goats, and boar.

The human lifestyle was apparently unchanged in the millennia that passed from the time of the Parpalló artists to that of the Levantines. But all indications are that Levantine art was produced by hunter-gatherers who were in contact with groups practicing a productive economy based on domesticated plants and animals. The artists themselves may have been those first Neolithic people who arrived there about 7,000 years ago. In any case, they represent the moment when one world was ending in this little corner of Iberia and another world, our world, was beginning. Ours is the world of the domesticated human, who moves at the weary pace of his or her cattle, and who bends to the earth to dig, looking up only to appeal for rain or to beg that it cease. The changed economy also produced ideological change. After all, the gods of the hunters were not the gods of the farmers and herders.

Someone wrote that when a language dies, those who spoke it in the past die a second time. The same can be said for the death of old myths and their replacement with new ones. We will never fully understand cave paintings because they do not speak to us. But something even more terrible has happened. The natural world also stopped speaking to humans. The stone arch of my childhood has ceased to be a bridge from the days of giants. The only sounds to be heard in the forest are the quarrying machinery in the entrails of the mountain, the factories where limestone is processed into cement, the chainsaws felling timber, and the passing of traffic on the highway.

The sun set on the free and untrammeled bands of hunters, on "uncivilized" humanity. Farmers arrived and with them comes the end of this book and my farewell to the reader, a farewell that I hope will be only temporary.

Afterword

Mauro Hernández, a scholar of Levantine art, tells me that vegetation and feral animals are returning to the ravines and precipices where rock shelters are found in the eastern peninsula. All around the paintings of the last Iberian hunters and gatherers, nature is reclaiming land once subdued by crops and livestock. New life is flourishing on formerly denuded soil. Perhaps a generation wiser than ours will one day again hear the voice of nature in the wind.

In Memoriam

Protest movements have arisen in recent years among indigenous peoples in various parts of the world, demanding that the bones of their ancestors held by museums be returned to them. Australian Aboriginals have demanded and received human fossils thousands of years old in order to return them to the places where they rested before scientists disturbed them. Although I understand the feelings of the Aboriginals, I think that the best way to honor our ancestors is to learn more about them. It is not easy to convince indigenous people that their ancestors are better off in protected chambers than in nature, although I hope that books like this one will contribute to a better understanding of science's value for all humankind.

Numerous human fossils have been unearthed by researchers like me, and more will be unearthed in the future. Many more, though, rest in the bosom of our Mother Earth. In a heartfelt tribute to them, I end with the words that the Romans used to close the epitaphs of their most beloved departed.

Sit tibi terra levis. May the earth rest lightly upon you.

Acknowledgments

In writing this book I have benefited from the opportunity to discuss the ideas presented herein with many colleagues. They are sure to remember our discussions when they read some of the pages within. The list would be very extensive, because I have learned from all of them. Nuria García, Ana Gracia, Carlos Lorenzo, Manuel Martín-Loeches, and Ignacio Martínez deserve special thanks because they read every draft of the book, and their valuable comments helped make it both clearer and more rigorous. My father, Pedro María Arsuaga Eguizábal, translated the narrative of Otto Herz about his expedition to retrieve the Beresovka mammoth, and he plotted the route on a map so that we could travel together in our imaginations to that cold and far-away Siberia.

Bibliography

Without pretension of providing an exhaustive bibliography for a field as broad as that of human evolution, I offer a list of general works, the majority published in recent years, and short lists of articles or books that provide additional information about the topics covered in individual chapters of this book. When they are translated I have opted for the Spanish versions.

General Books

Aiello, L., and Dean, C., *An Introduction to Human Evolutionary Anatomy*, Academic Press, San Diego, 1990.

Anguita, F., and others, *Origen y evolución. Desde el Big Bang a las sociedades complejas*, Fundación Marcelino Botín, Santander, 1999.

Ayala, F.J., *Origen y evolución del hombre*, Alianza Editorial, Madrid, 1980.

Day, M., *Guide to Fossil Man*, Cassell, London, 1986.

Carbonell, E., Bermúdez de Castro, J. M.., Arsuaga, J. L., and Rodriguez, X. P. (eds.), *Los primeros pobladores de Europa: últimos descubrimientos y debate actual*, Diario de Burgos y Caja de Burgos, Burgos, 1998.

Cerdeño, M. L., and Vega, G., *La España de Altamira. Prehistoria de la Península Ibérica*, Historia 16, Madrid, 1995.

Conroy, G., *Primate Evolution*, W. W. Norton, New York, 1990.

Fleagle, J., *Primate Adaptation and Evolution*, Academic Press, San Diego, 1988.

Foley, R., *Another Unique Species*, Longman, Harlow, 1987.

Johanson, D., and Edey, M., *El primer antepasado del hombre*, Planeta, Barcelona, 1982.

Johanson, D., and Edgar, B., *From Lucy to Language*, Simon & Schuster, New York, 1996.

Johanson, D., Johanson, L., and Edgar, B., *Ancestors. In Search of Human Origins*, Villard Books, New York, 1994.

Johanson, D., and Shreeve, J., *Lucy's Child*, Morrow, New York, 1982.

Jones, S., Martin, R., and Pilbeam, D. (eds.), *The Cambridge Encyclopedia of Human Evolution*, Cambridge University Press, Cambridge, 1992.

Klein, R., *The Human Career*, University of Chicago Press, Chicago, 1989.

Le Gros Clark, W., *The Antecedents of Man*, Quadrangle, Chicago, 1969.

Leakey, R., *La formación de la humanidad*, Serbal, Barcelona, 1981.

Leakey, R., and Lewin, R., *Nuestras orígines*, Crítica, Barcelona, 1994.

Lewin, R., *La interpretación de los fósiles*, Planeta, Barcelona, 1989.

———, *Evolución humana*, Salvat, Barcelona, 1994.

Martin, R., *Primate Origins and Evolution*, Chapman & Hall, London, 1990.

Moure, A. (ed.), *"El hombre fósil" 80 años después*, University of Cantabria, Santander, 1996.

Moure, A., *El origen del hombre*, Historia 16, Madrid, 1997.

Napier, J., and Napier, P., *The Natural History of Primates*, British Museum (Natural History), London, 1985.

Reader, J., *Eslabones perdidos*, Fondo Educativo Interamericano, México, 1982.

Rightmire, G., *The Evolution of* Homo erectus, Cambridge University Press, Cambridge, 1990.

Stringer, C., and Gamble, C., *En busca de los neandertales*, Crítica, Barcelona, 1996.

Stringer, C., and McKie, R., *African Exodus*, Jonatham Cape, London, 1996.

Szalay, F., and Delson, E., *Evolutionary History of the Primates*, Academic Press, New York, 1995.

Tattersall, I., Delson, E., and Van Couvering, J. (eds.), *Encyclopedia of Human Evolution and Prehistory*, Garland, New York, 1988.

Trinkaus, E., and Shipman, P., *The Neandertals*, Vintage, New York, 1992.

Walker, A., and Shipman, P., *The Wisdom of the Bones*, Knopf, New York, 1996.

Chapters 1 and 2: The Solitary Species and The Human Paradox

Abbate, E., and others, "A One Million Year Old *Homo* Cranium from Danakil (Afar) Depression of Eritrea," *Nature* (1998), vol. 393, pp. 458-460.

Asfaw, B., White, T., Lovejoy, O., Latimer, B., Simpson, S., and Suwa, G., "*Australopithecus garhi*. A New Species of Early Hominid from Ethiopia," *Science* (1999), vol. 284, pp. 629-635.

Conroy, G., Weber, G., Seidler, H., Tobias, P., and Kane, A., "Endocranial Capacity in an Early Hominid Cranium from Sterkfontein, South Africa," *Science* (1998), vol. 280, 1.730-1.731.

Clarke, R., "First Ever Discovery of a Well-Preserved Skull and Associated Skeleton of *Australopithecus*," *South African Journal of Science* (1998), vol. 94, pp. 460-463

DeHeinzelin, J., Clark, D., White, T., Hart, W., Renne, P., WoldeGabriel, G., Beyene, Y., and Vrba, E., "Environment and Behavior of 2.5 Million Year Old Bouri Hominids," *Science* (1999), vol. 284, pp. 625-629.

DeMenocal, P., "Plio-Pleistocene African Climate," *Science* (1995), vol. 270, pp. 53-59

Dubar, R., "The Social Brain Hypothesis," *Evolutionary Anthropology* (1998), vol. 6, pp. 178-190.

Falk, D., Froese, N., Stone, D., and Dudek, B., "Sex Differences in Brain/Body Relationships of *Rhesus* Monkeys and Humans," *Journal of Human Evolution* (1999), vol. 36, pp. 233-238.

Leakey, M., Feibel, C., McDougall, I., Ward, C., and Walker, A., "New Specimens and Confirmation of an Early Age for *Australopithecus anamesis*," *Nature* (1998), vol. 393, pp. 62-66.

Otte, M., "Naissance des formes," *Melanges Pierre Colman* (1996), vol. 15, pp. 14-17.

Walker, A., and Shipman, P., *The Wisdom of Bones*, Alfred A. Knopf, New York, 1996.

Chapter 3: The Neanderthals

Arsuaga, J. L., Martínez, I., Gracia, A., and Lorenzo, C., "The Sima de los Huesos Crania (Sierra de Atapuerca, Spain). A Comparative Study," *Journal of Human Evolution* (1997), pp. 219-281.

Braüer, G., Yokohama, Y., Falguéres, C., and Mbua, E., "Modern Human Origins Backdated," *Nature* (1997), vol. 386, p. 337.

Coon, C., *Adaptaciones raciales*, Labor, Barcelona, 1984.

Chen T., Yang, Q., and Wu, E., "Antiquity of *Homo sapiens* in China," *Nature* (1994), vol. 368, pp. 55-56.

Churchill, S., "Cold Adaptation, Heterochrony, and Neandertals," *Evolutionary Anthropology* (1998), vol. 7, pp. 46-61.

Formicola, V., and Giannecchini, M., "Evolutionary Trends of Stature in Upper Paleolithic and Mesolithic Europe," *Journal of Human Evolution* (1999), vol. 36, pp. 319-333.

Rak, Y., "The Neanderthal: A New Look to an Old Face," *Journal of Human Evolution* (1999), vol. 15, pp. 151-164.

Ruff, C., Trinkaus, E., and Holliday, T., "Body Mass and Encephalization in Pleistocene *Homo*," *Nature* (1997), vol. 387, pp. 173-176.

Ruff, C., and Walker, A., "Body Size and Body Shape," in Walker, A., and Leakey, R. (eds.), *The Nariokotome* Homo erectus *Skeleton* (1993), Harvard University Press, Cambridge, pp. 234-265.

Trinkaus, E., "The Neandertal Face: Evolutionary and Functional Perspectives on a Recent Hominid Face," *Journal of Human Evolution* (1986), vol. 16, pp. 429-443.

Chapters 4 and 5: The Animated Forest and The Reindeer are Coming!

Aguirre, E. (ed.), *Atapuerca y la evolución humana*, Fundación Ramón Areces, Madrid, 1998.

Alcolea, J. J., Balbín, R. de, García, M. A., and Jiménez, P. J., "Nouvelles découvertes d'art rupestre paléolithique dans le centre de la Péninsule ibérique: la grotte du Renne (Valdestos, Guadalajara)," *L'Anthropologie* (1997), vol. 101, pp. 144-163.

Altuna, J., "Le Paléolithique moyen de la région cantabrique," *L'Anthropologie* (1992), vol. 96, pp. 87-102.

———, "Faunas de clima frío en la Península Ibérica durante el Pleistoceno superior," in Ramil-Rego, P., Fernández, C., and Rodríguez, M. (eds.), *Biogeografía Pleistocena-Holocena de La Península Ibérica*, Xunta de Galicia, Santiago de Compostela, 1996, pp. 13-42.

Bermúdez de Castro, J. M., Arsuaga, J. L., and Carbonell, E. (eds.), *Evolución humana en Europa y los yacimientos de la Sierra de Atapuerca*, Junta Castilla y León, Consejería de Cultura y Turismo, Valladolid, 1992.

Blanco, E., and others, *Los bosques ibéricos*, Planeta, Barcelona, 1997.

Castañón, J. C., and Frochoso, M., "Hugo Obermaier y el glaciarismo pleistoceno," in A. Moure (ed.) *"El Hombre Fósil" 80 años después*, University of Cantabria, Santander, 1996, pp. 153-175.

Corchón, S., "La corniche cantabrique entre 15000 et 13000 ans BP: la perspective donnée par l'art mobilier," *L'Anthropologie* (1997), vol. 101, pp. 114-143.

Chapa, T., and Menéndez, M. (eds.), *Arte Paleolítico*, Editorial Complutense, Madrid, 1994.

Dupré, M., *Palinología y paleoambiente*, Servicio de Investigación Prehistórica, Valencia, 1988.

Ferreras, C., and Arozena, M. A., *Los bosques*, Alianza Editorial, Madrid, 1987.

García, N., and Arsuaga, J. L., "The Carnivore Remains from the Hominid-Bearing Trinchera-Galería, Sierra de Atapuerca, Middle Pleistocene Site (Spain)," *Geobios* (1998), vol. 31, pp. 659-674.

Garci, M., *El oeste de Europa y la Península Ibérica desde hace—120,000 años hasta el presente*, Enresa, Madrid, 1994.

Guérin, C., and Patou-Mathis, M., *Les grands mammifères Plio-Pléistocénes d'Europe*, Masson, Paris, 1996.

Gutiérrez Elorza, M. (ed.), *Geomorfología de España*, Rueda, Madrid, 1994.

Hertz, O., *Ausgrabung eines Mamuthkadavers*, Academia Imperial de Ciencias, St. Petersburg, 1902.

Kahlke, H. D., *Die Eiszeit*, Die Deutsche Bibliothek, Jena, 1994.

Kahlke, R., "Die Entstehungs–, Entwicklungs– und verbreitungsgeschichte des oberpleistozänen *Mammuthus-Coelodonta*-Faunenkomplexes in Eurasien (Grossäuger)," *Abhandlungen der senckenergischen naturforschenden Gesellchaft 546* (1994), pp. 1-164.

Kurtén, B., *The Cave Bear Story. Life and death of a Vanished Animal*, Columbia University Press, New York, 1976.

Pedraza, J. de, Carrasco, R. M., and Díez-Herrero, A., "Morfoe-structura y modelado del Sistema Central español," in Segura, M., De Bustamante, I., and Bardají T. (eds.), *Itinerarios geológicas desde Alcalá de Henares*, University of Alcalá, Alcalá de Henares, 1996, pp. 55-80.

Turner, A., and Antón, M., *The Big Cats and their Fossil Relatives*, Columbia, New York, 1997.

Van der Made, J., "Ungulates from Gran Dolina (Atapuerca, Burgos, Spain)," *Quaternaire* (1998), vol. 9, pp. 267-281.

Chapter 6: The Grand Extinction

Álvarez, B. T., "Plantas tóxicas usadas para percar en nuestros ríos," *Quercus* (1998), vol. 147, pp. 36-37.

Auguste, P., "Érude archéozoologique des grands mammiféres du site pléistocéne moyen de Biache-Saint-Vaast (Pas-de-Calais, France): apports biostratagraphiques et palethnographiques," *L'Anthropologie* (1992), vol. 96, pp. 49-69.

Caussimont, G., and Hartasánchez, R., "El oso pardo y las estaciones," *Quercus* (1996), vol. 119, pp. 31-37.

Hawkes, K., O'Connell, J., and Blurton Jones, N., "Hadza Women's Time Allocation, Offspring Provisioning, and the Evolution of Long Postmenopausal Life Spans," *Current Anthropology* (1997), vol. 38, pp. 551-577.

Howell, F. C., "The Evolution of Human Huntings," *Journal of Human Evolution* (1989), vol. 18, pp. 583-594.

Birket-Smith, K., *Los esquimales*, Labor, Barcelona, 1965.

López-Piñero, J. M., "El megaterio," *La Aventura de la Historia* (1999), vol. 3, pp. 88-89.

Llaneza, L., "Hábitos alimenticios del lobo en la cordillera Cantábrica," *Quercus* (1999), vol. 157, pp. 16-19.

Marean, C., "A Critique of the Evidence for Scavenging by Neandertals and Early Modern Humans: New Data from Kobeh Cave (Zagros Mountains, Iran) and Die Kelders Cave 1 Layer 10 (South Africa)," *Journal of Human Evolution* (1998), vol. 35, pp. 111-136.

Monchot, H., "La caza del muflón (*Ovis Antiqua* Pommerol, 1879) en el Pleisteceno Medio de los Pirineos: el ejemplo de la cueva de l'Aragó (Tautavel, France)," *Revista Española de Paleontología* (1999), vol. 14, pp. 67-78.

Notario, R., *El oso pardo en España*, Ministerio de Agricultura, Madrid, 1970.

O'Connell, J., Hawkes, K., and Blurton Jones, N. G., "Hadza Scavenging: Implications for Plio-Pleistocene Hominids Subsistence," *Current Anthropology* (1988), vol. 29, pp. 356-363.

Pérez-Pérez, A., Bermúdez de Castro, J. M., and Arsuaga, J. L., "Nonocclusal Dental Microwear Analysis of 300,000 Year Old *Homo heidelbergensis* Teeth from Sima de los Huesos (Sierra de Atapuerca, Spain)," *American Journal of Physical Anthropology* (1999), vol. 108, pp. 433-457.

Santonja, M., López Martínez, N., and Pérez-González, A. (eds.), *Ocupaciones achelenses en el valle del Jarama (Arganda, Madrid)*, Diputación Provincial de Madrid, Madrid, 1980.

Schoeninger, M., "Stable Isotope Studies in Human Evolution," *Evolutionary Anthropology* (1995), vol. 3, pp. 83-98.

Villa, P., "Torralba and Áridos: Elephant Exploitation in Middle Pleistocene Spain," *Journal of Human Evolution* (1990), vol. 19, pp. 299-309.

Chapter 7: A Poisoned Gift

Arsuaga, J. L., Martínez, I., Gracia, A., Carretero, J. M., Lorenzo, C., García, N., and Ortega, A. I., "Sima de los Huesos (Sierra de Atapuerca, Spain). The Site," *Journal of Human Evolution* (1997), vol. 33, pp. 109-127.

Bentley, G., "Aping Our Ancestors: Comparative Aspects of Reproductive Ecology," *Evolutionary Anthropology* (1999), vol. 7, pp. 175-185.

Bermúdez de Castro, J. M., and Nicolás, E., "Paleodemography of the Atapuerca-SH Middle Pleistocene Hominid Sample," *Journal of Human Evolution* (1997), vol. 33, pp. 333-335.

Bocquet-Appel, J. P., "Small Populations: Demography and Paleoanthropological Inferences," *Journal of Human Evolution* (1985), vol. 14, pp. 683-691.

Bocquet-Appel, J. P., and Arsuaga, J. L., "Age Distributions of Hominid Samples at Atapuerca (SH) and Krapina Could Indicate Accumulation by Catastrophe," *Journal of Archaeological Science* (1999), vol. 26, pp, 327-338.

Blurton Jones, N., Smith, L., O'Connell, J., Hawkes, K., and Kamuzora, C., "Demography of the Hadza, an Increasing and High Density Population of Savanna Foragers," *American Journal of Physical Anthropology* (1992), vol. 89, pp. 159-181.

Goodall, J., *A través de la ventana*, Salvat, Barcelona, 1993.

Hill, K., and Hurtado, M., *Ache Life History*, Aldine de Gruyter, New York, 1996.

Howell, N., *Demography of the Dobe !Kung*, Academic Press, New York, 1979.

Trinkaus, E., "Neanderthal Mortality Patterns," *Journal of Archaeological Science* (1999), vol. 157, pp. 121-142.

Vicente, J. L., Rodríguez, M., and Palacios, J., "Relaciones entre lobos y ciervos en la sierra de la Culebra," *Quercus* (1999), vol. 157, pp. 10-15.

Chapter 8: Children of the Fire

Crick, F., *La búsqueda científica del alma*, Círculo de Lectores, Barcelona, 1994.

Duff, A., Clark, G., and Chadderon, T., "Symbolism in the Early Paleolithic: A Conceptual Odyssey," *Cambridge Archaeological Journal* (1992), vol. 2, pp. 211-229.

Deacon, T., *The Symbolic Species*, The Penguin Press, Harmondsworth, Middlesex, 1997.

Falk, D., *Braindance*, Henry Holt, New York, 1992.

Lieberman, P., *Eve Spoke*, Picador, London, 1998.

Macphail, E., *The Evolution of Consciousness*, Oxford University Press, Oxford, 1998.

Mithen, S., *Arqueología de la mente*, Crítica, Barcelona, 1998.

Mosterín, J., *¡Vivan los animales!*, Temas de Debate, Madrid, 1998.

Noble, W., and Davidson, I., *Human Evolution, Language and Mind*, Cambridge University Press, Cambridge, 1996.

Tattersall, I., *The Origin of the Human Capacity*, 68th James Arthur Lecture, American Museum of Natural History, New York, 1998.

———, *Hacia el ser humano*, Península, Barcelona, 1998.

Weiner, S., Xu, Q., Goldberg, P., Liu, J., and Bar-Yosef, O., "Evidence for the Use of Fire at Zhoukadian, China," *Science* (1998), vol. 281, pp. 251-253.

Wu, X., "Investigating the Possible Use of Fire at Zhoukadian, China," *Science* (1999), Vol. 283, p. 299.

Chapter 9: And the World Was Made Transparent

Appenzeller, T., Clery D., and Culotta, E., "Archaeology: Transitions in Prehistory," *Science* (1998), vol. 282, 1.441-1.458.

D'Errico, F., Zilhão, J., Julien, M., Baffier, D., and Pelegrin, J., "Neanderthal Acculturation in Western Europe," *Current Anthropology* (1998), vol. 39, pp. 1-44.

Gamble, C., "Gibralter and the Neanderthals 1848-1998," *Journal of Human Evolution* (1996), vol. 36, pp. 239-243.

Hoffecker, J. F., "Neanderthals and Modern Humans in Eastern Europe," *Evolutionary Anthropology* (1998), vol. 7, pp.129-141.

Mellars, P., "The Fate of Neanderthals," *Nature* (1998), vol. 395, pp. 539-540.

Pike-Tay, A., Cabrera, V., and Bernaldo de Quirós, F., "Seasonal Variations of the Middle-Upper Paleolithic Transition at El Castillo, Cueva Morín and El Pendo (Cantabria, Spain)," *Journal of Human Evolution* (1999), vol. 36, pp. 283-317.

Strauss, L. G., "The Upper Paloelithic of Europe: An Overview," *Evolutionary Anthropology* (1995), vol. 4, pp. 4-16.

———, "The Iberian Situation Between 40,000 and 30,000 B.P. in Light of European Models of Migration and Convergence," in Clark, G. A., and Willerment, C.M. (eds), *Conceptual Issues in Modern Human Origins Research*, Aldine de Gruyter, New York, 1997, pp. 235-252.

Taborin, Y., "L'art des premiéres parures," in Sacco, F., and Sauvet, G. (eds.), *La propre de l'homme. Psychoanalyse et préhistoire*, Delachaux et Niestlé, Lausana, 1998, pp. 123-150.

Van Andel, T., "Middle and Upper Paleolithic Environments and the Calibration of 14C Dates Beyond 10,000 B.P.," *Antiquity* (1998), vol. 72, pp. 26-33.

Vega-Toscano, G., Hoyas, M., Ruiz-Bustos, A., and Laville, H., "La séquence de la Grotte de la Carihuela (Piñar, Grenade): Chronostratrigraphie et paléoécologie de Pléistocéne supérieur au sud de la Péninsule Ibérique," in Otte, M. (ed.), *L'Homme de Néandertal. L'Environnement*, University of Liége, Liége, 1998, pp. 169-180.

Epilogue

Hernández, M., Ferrer, P., and Catalá, E., *Arte rupestre en Alicante*, Fundación Banco Exterior, Alicante, 1988.

———, *L'Art llevantí*, Centre d'Estudis Contestans, Cocentaina, 1998.

Villaverde, V., *Arte paleolítico de la Cova del Parpalló*, Diputació de Valéncia, Valencia, 1994.

Index

Discover the triumphs in human evolution – from desire to DNA.

0470 85144 9 • £16.99

An original theory on human development, nature and the future that suggests that it is our desire to quest – for food and shelter, for knowledge, for wealth, for adventure – coupled with our unique physical abilities to do so that have controlled our evolution and have led humans to develop away from closely related animals.

0470 85429 4 • £18.99

"...As a biographer McElheny presents the science in this amazing story with effortless lucidity and sets out clearly Watson's success as scientific impresario...McElheny has done a good job, producing a warts-and-all portrait..."

The Guardian